Economic Instruments for Water Management

The Cases of France, Mexico and Brazil

Ronaldo Seroa da Motta

Professor of Environmental Economics and Co-ordinator of Regulation Studies at the Research Institute for Applied Economics (IPEA), Brazil

Alban Thomas

Director of Research at the French Institute for Research in Agriculture (INRA), Lecturer in Econometrics and Environmental Economics, University of Toulouse, France

Lilian Saade Hazin

Independent Consultant in the Water Sector

José Gustavo Feres

Researcher, Research Institute for Applied Economics (IPEA), Brazil

Céline Nauges

Researcher, French Institute for Research in Agriculture (INRA), France

Antonio Saade Hazin

Independent Consultant on Public Policy

Edward Elgar

Cheltenham, UK • Northampton MA, USA

Published by
Edward Elgar Publishing Limited
Glensanda House
Montpellier Parade
Cheltenham
Glos GL50 1UA
UK

Edward Elgar Publishing, Inc.
136 West Street
Suite 202
Northampton
Massachusetts 01060
USA

A catalogue record for this book
is available from the British Library

ISBN 1 84376 964 6

Printed and bound in Great Britain by MPG Books Ltd, Bodmin, Cornwall

Contents

v

Figures

Tables

x   *Tables*

1. Introduction[1]

Ronaldo Seroa da Motta

The economic literature commonly identifies economic instruments (EIs) as a 'better' way to achieve environmental goals than specified quantity and technological standards commonly known as command-and-control mechanisms (CACs). However, the choice of an appropriate economic instrument is theoretically complex; and the experiences with their application are full of controversy about their effectiveness in accomplishing desired environmental targets.

Environmental regulators usually apply two different kinds of individual standards to induce agents to conform to environmental or resource management goals. One type specifies emission rates or levels, as well as permissible rates of resource (water) use. The other kind of standard is technologically defined, requiring for example a specific kind of pollution control technology. Overall ambient standards also are set to identify the aggregate environmental performance targets in, for example, a watershed. Theoretically, individual technology or performance standards would be set in order to make agents achieve, in aggregate, the ambient standards.

If instead of complying with a uniform individual standard, pollution or resource use levels were charged, each agent's control or use level would be different. Agents with lower costs will control more than agents with higher costs (and agents with higher demand responsiveness for the water resource will reduce consumption less, for a given charge, than those with a lower demand responsiveness). With a uniform pollution charge, all private agents will choose between paying the charge and reducing the effluent to the point where they have the same marginal cost of control. In this situation, control will then follow the least-cost path among agents for a given reduction in aggregate pollution loading. In other words the reduction in aggregate loading will be cost-effective.

Equity issues also can become more interesting than with a single standard for everyone. Since each source's control level will be

1

determined by the point that the pollution or use price equals marginal control cost, by definition all controlled units will cost less than the charged price whereas pollution and use level not controlled will be charged at this pollution/use price.

In addition to that, charging all units of pollution and use will create a stronger dynamic incentive for eco-friendly technological change (resource conservation, pollution prevention or treatment), since all remaining units of pollution or water use will incur a cost, not just those required to achieve compliance status. In most cases, however, the ambient environmental consequences of the loading reduction from a uniform pollution charge will depend on where the reductions occur as well as by how much.

While a uniform charge will yield cost-effectiveness gains compared to the same reduction in loading through uniform individual standards, the environmental performance of such pricing mechanisms depends on whatever ambient environmental standards the charge system seeks to achieve. If these underlying environmental standards are weak or fail to capture all relevant ecological dimensions, then the pricing mechanisms also will be limited in their environmental consequences. Nevertheless pricing with these limitations may still offer the least-cost approach to total pollution reduction when compared to standards with the same ecological restrictions. As already noted, the overall environmental performance of both charges and standards can depend on where the reductions occur, as well as on how much. The more homogeneous the effects of pollution from different sources on ambient conditions, the less important this consideration will be.

Although theoretically pricing instruments promise cost-effectiveness relative to control instruments, the potential cost savings with economic instruments will depend on the degree of control of cost heterogeneity among polluters and users, which depends in turn on size, technology, information, managerial skills and other factors. If marginal costs of control are not too different, then the cost savings from the charge system will be more modest. The overall cost-effectiveness of a charge system also depends on the costs of obtaining all the required information and to set up the charging and monitoring basis. The system may have high transaction costs that could dissipate the expected cost savings. Therefore regulators must have in mind that control cost savings must be balanced against implementation costs including staffing and monitoring facilities.

Seroa da Motta et al. (1999) first presented a comprehensive survey of Latin America and the Caribbean's EI experiences in the mid-1990s. They concluded that there is a wide range of application of EIs in the region and that they have closely followed the OECD pattern with revenue-raising aims. Water charges have been the most advanced case. They have, however, showed problems in design and implementation issues, such as:

- weak targeting and performance monitoring of environmental goals
- lack of sound pricing criteria
- poor performance on revenue collection

Among experiences in developed countries, the French river basin system has been seen as a paradigm for Latin America experiences. This was mostly due to the fact that the French system was created by governmental decision quite recently and implemented in a reasonable time with immediate results. Praising of this experience has obscured the identification of its main difficulties and constraints that, once recognized could be of great value for followers, particularly when countries in the region are already struggling to initiate or improve the implementation of their systems.

Based on this diagnostic, after presenting a summary review of the literature, this volume reviews water charge experiences in France, Mexico and Brazil. These country reviews were undertaken by local experts and, despite differences in emphasis, each country evaluation is organized along the same guidelines covering topics of relevance for the application of a policy instrument, namely:

1. Policy analysis phase: the policy setting in which the water EI was introduced as a mean of achieving policy goals.
2. Instrument design phase: the theoretical, institutional and legal basis on which the EI was conceived.
3. Instrument implementation phase: successes and failures of the EI application and its review process.

1.1   COUNTRY CASES

The 1964 Water Act profoundly modified the French water management system and its apparent success later set a new paradigm for water policies across the world, particularly in Latin America.

4   *Economic instruments for water management*

The new approach in France was based on two general principles: decentralization and planning. Decentralization is based on the idea that water management organization should reflect the physical unity of water bodies in order to account for the potential sources of conflicts. To handle the externality problems linked to water pollution and conflicts of use as an integrated approach, the river basin is defined as the basic administrative unit rather than addressing uses differently and water management centrally with one set of unified performance standards. Planning is intended to provide consistent decisions at the river basin level and to introduce a medium-term perspective on water management.

The decentralization principle is put into practice by the creation of Water Agencies and River Basin Committees in each of the six French river basins. While the former are intended to perform executive functions, River Basin Committees act as consultative bodies. To carry this on, two new instruments were adopted: five-year management plans and water charges.

Implementation of water charges was gradual and it worked very well to generate revenues for water-related investments, with much of the revenue transferred back towards water charge payers. However, no major role for price incentives has been found at the prevailing water charge levels to induce changed water use patterns. Moreover the special treatment to agricultural users through exemptions has led to the sacrifice of noticeable environmental gains.

Close to the French approach, Mexico has adopted water charges since the 1980s. The water use charge from federal water bodies has been in place since 1986 and the wastewater charge since 1991. A central agency in charge of the use of federal water resources, the National Water Commission (CNA), was created in 1989. The CNA is the sole authority for federal water management and is responsible for the promotion and execution of federal infrastructure and the necessary services for the preservation of water quality. The most recent regulation is the 1992 National Water Law that is the backbone of the federal water system.

The CNA is attached to the Ministry of Environment and Natural Resources. The federal water management system encompasses 13 administrative regions defined by the CNA, following hydrographic criteria. Each region comprises one or more basins, thus basins and not states are the basic division of the Mexican water management system. In total the system includes 26 Basin Councils. Following the French principles, the objective of the councils is to promote participation in the management process of the basin. Pricing criteria of water charges are, however, set by a federal law revised every year. These councils moreover have not been fully implemented and their capability has not been completely developed in order that they may be fully evaluated.

Water pollution charges in Mexico perform the role of a non-compliance charge since polluters only pay for units above the discharge standard. However, the implementation of the water charges has not been very successful since national coverage of the vast country's water system has required monitoring resources and enforcement capability beyond the CNA institutional capacity. In addition to that, CNA institutional power has been more concerned with infrastructure development than pursuing environmental targets. The reduced scope for private and public participation, associated with lack of information based on careful analysis of expected impacts from charge incidence has created polluters' opposition on competitiveness and distributive grounds.

As a consequence of these institutional barriers, revenue generation has been very low and no changes in water use pattern have occurred. A project law entitled Ley de Cuencas y Aguas Nacionales (Basin and National Water Law) and at least two other projects of reform to the current law are being analysed by the Congress. Although they are still in the discussion phase, it is worth mentioning that they all propose giving more autonomy to the river basin institutions. While the autonomy proposed might not go as far as it could, the undeniable goal of every proposal is to strengthen the institutional capacity of the participatory institutions.

The Brazilian experience is quite different. Following the approval of the Federal Water Law (Law 9433 of 1997), Brazil has recently implemented a wide-ranging water sector reform, including the introduction of environmental water charges. The Brazilian legal framework for water resources management is based on the constitutional distinction between federal and state waters. Federal waters are those that flow across state boundaries or along the boundaries between two or more states or a foreign country. State waters are those situated entirely within the territory of a single state.

The new water management system adopted the same French principles of management by Water Basin Committees and Agencies where water charges are associated with River Basin Management Plans that identify environmental targets to be accomplished with a set of water-related investments and financed with water charge revenues. However, pricing criteria for the setting of charges have no general structure and committees have more autonomy in this matter than in France.

The creation of River Basin Committees is also less centralized. It depends on the users' initiatives to form a committee that must fulfil some managerial requirements and then must seek approval of the National Council of Water Resources. So the river basin national grid will be gradually implemented. The National Council of Water Resources also deals with inter-basin and inter-state disputes. Supervising and assisting basin water agencies is the National Water Agency.

The first implementation of water charges in federal rivers is due to begin in March 2003 in the Paraíba do Sul river basin, where a single low charge will be levied on users for only a small number of pollutants. In this initial phase, the aim is to collect enough charge revenue to entitle the basin to compete for federal funds oriented toward water clean-up projects.

At state levels, almost all states have their own water policy based on the principles adopted in the national framework. Ceará already has water use charges and São Paulo is also near to implementing its charge system very similar to the one adopted for the Paraíba do Sul river basin.

There is no doubt that, so far, Brazilian experiences have followed revenue-raising aims as in France and the participatory process through River Basin Committees was also not able to introduce clear price incentives for changes in water use patterns.

1.2   MAIN CONCLUSIONS

Water charges have been introduced as instruments for an integrated water policy approach. This approach has been primarily concerned with (1) the need to plan and decentralize water management in order to accommodate multiple conflicting uses and excesses over assimilative and support capacities of the country's water systems, and (2) the need to raise revenue.

Despite the fact that the primary goal of water charges has been in principle to assign an economic value for water, in all cases charges were in place mainly to support the achievement of CAC instruments, such as discharge permits and standards.

Decentralization was planned in two ways: (1) water management goals and targets differentiated by river basins, and (2) conflicts among users dealt with through a participatory process. The main institutional bases has been the River Basin Committees that define management targets to be executed by their Water Agencies.

To accommodate economic and social conflicts, water charges in practice are financing mechanisms for investment solutions for water management, including pollution control investments.

Apart from administrative costs, the major share of water charge revenues goes to infrastructure investments and direct transfer for users to finance their pollution abatement actions. Such transfers are thought of as the cornerstone for political acceptance and users' commitment to the charge system since sectoral interests reduce the possibility of fully applied water charges.

The need for a participatory process to accommodate users' conflicts and to increase acceptance reduces the potential benefits of a water charge system. That is, participation may solve revenue-related conflicts but it does not necessarily create incentives for a charge system that will significantly change water pattern uses.

It is also recognized that the lack of a continuous evaluation process to analyse the effects of the charge system on use levels and on environment quality has delayed improvements in the system and in the allocation of the water charge revenues.

1.3 RECOMMENDATIONS

Recommendations can be summarized as follows:

- Revenue-raising goals should be explicitly acknowledged and achieving environmental goals has to be planned. But it is important to make explicit the environmental consequences of charge application to allow for gradual incorporation of environmental criteria in the charge system.
- Autonomy of river basin authorities must be tailored to maximize institutional capacity by facilitating political acceptance, reducing information gaps and reducing administrative costs.
- Also the water management framework must be integrated with other policy frameworks to increase monitoring and enforcement capacities.
- Continuous environmental evaluation of the river basin should be

undertaken incorporating economic models that identify water use changes related to charge impacts. And cost–benefit analytical tools should be developed for projects to be financed with charge revenues to maximize the social value of the investment actions.

● Public opinion should be brought into the debate on water management issues with data release and technical arguments to consolidate river basin management and the role of water charges.

NOTE

1. This chapter was part of a series of papers commissioned by the Inter-American Development Bank for the Environmental Policy Dialogue and the opinions expressed in this chapter are solely those of the author and do not necessarily reflect the position of the IADB.

BIBLIOGRAPHY

Acquatella, J. (2001), 'CEPAL Aplicación de Instrumentos Económicos en la Gestión Ambiental en América Latina y el Caribe: Desafíos y Factores Condicionantes', Serie Medio Ambiente y Desarrollo No. 31, Santiago: CEPAL.

Bovenberg, A.L. and L.H. Goulder (1996), 'Optimal environmental taxation in the presence of other taxes: general equilibrium analyses', *American Economic Review*, **86**, 985–1000.

CAEMA (2001), 'Evaluación de la efectividad ambiental y eficiencia económica de la tasa por contaminación hídrica en el sector industrial colombiano', El Centro Andino para la Economía en el Medio Ambiente, Bogotá: mimeo.

Fullerton, D. (1997), 'Environmental levies and distortionary taxation: comment', *American Economic Review*, **87**, 245–51.

OECD (1994), *Managing the Environment: The Role of Economic Instruments*, Paris: OECD.

OECD (1995), *Environmental Taxes in OECD Countries*, Paris: OECD.

Parry, I.W.H., R.C. Williams III and L.H. Goulder (1999), 'When can carbon abatement policies increase welfare? The fundamental role of distorted factor markets', *Journal of Environmental Economics and Management*, **37** (1), 52–84.

Seroa da Motta, R. (2001), 'Tributación ambiental, macroeconomia y médio ambiente en América Latina: aspectos conceptuales y el caso de Brasil', Serie Macroeconomía del Desarrollo 7, Santiago: CEPAL.

Seroa da Motta, R., R. Huber and J. Ruintenbeek (1999), 'Market based instruments for environmental policymaking in Latin America and the Caribbean: lessons from eleven countries', *Environment and Development Economics*, **4** (2), 177–202.

2. Conceptual framework and review of experiences[1]

Ronaldo Seroa da Motta

2.1 INTRODUCTION

The use of economic instruments (EI) has been analysed in a vast literature. Theoretical analysis started with the pioneer proposal of externality taxation of Pigou and led to the detailed and comprehensive theoretical work of Baumol and Oates (1988). Add to this numerous books and articles dealing with specific particularities in distinct contexts. The same abundance is found within the literature of EI experiences. It is specially worth noting the diverse OECD (for example, 1994, 1995) publications that describe and analyse every single and specific case of EI in developed countries. A detailed survey for selected developing countries can be found in Rietbergen-McCracken and Abaza (2000). A recent survey of Stavins (2002) covers all regions in the world, pointing out reasons for failures and successes. For Latin America and the Caribbean the World Bank research in Seroa da Motta et al. (1999) presents the first comprehensive analysis of EI uses, later complemented by a Cepal study in Acquatella (2001). A useful earlier citation is Panayotou (1993).

This chapter will not attempt to propose new theoretical or analytical insights relevant to EI use in the Latin American region. Instead it will point out the issues raised in the literature that will be most useful to understanding and evaluating the case studies of water EIs that are fully described in the following chapters. In doing so, it will not only present these theoretical and practical issues but it will also discuss the existing experiences that will help to elucidate the discussion.

We will not address other important pricing EIs, such as deposit-return schemes for solid wastes and demand-oriented schemes such as eco-labelling. Although they can be ancillary instruments for water use and pollution, they do not fit directly into water management practices that will be analysed.

Mixing theoretical and conceptual notes with a brief review of practical cases we also hope to offer a framework that clarifies the most important issues of EI design and application, so helping future initiatives. Along these lines, at the end of the chapter, we have added suggested guidelines for EI formulation.

2.2   CONCEPTUAL FRAMEWORK

The basic principle of economic instruments is the 'polluter/user pays principle' that shifts the initial costs of uses of natural resources from society as a whole to polluters and users.

Any environmental regulation, such as norms, standards, taxes, quotas and so on, poses such costs on polluters and users, and therefore changes relative prices of natural resources. Consequently there will be economic incentives to alter pollution and use patterns and so any of those instruments can be considered as economic instruments.

According to Seroa da Motta et al. (1999), no standardized definition of an EI exists. A 'weak' EI uses regulations and is usually denominated as command-and-control (CAC) in the literature, whereas a 'strong' EI uses mainly market forces to decentralize decision-making and is commonly referred to as typical market-based instruments.

The 'strength' of an EI depends on the degree of flexibility that a polluter has in achieving a given environmental target. A 'weak' EI uses regulation to dictate the type of process that must be used, and failure to comply results in economic sanctions. A 'strong' EI reckons mainly on market forces to determine the best way to meet a given standard or goal.

'Flexibility' refers to the degree to which social (or state) decisions are transferred to the private (individual) level. A strong EI decentralizes decision-making, giving the polluter or resource users a maximum amount of flexibility to select the production or consumption options that minimizes its control, thus ensuring that profit- or utility-maximizing behaviour generates the 'lowest social cost' outcome of achieving a particular level of environmental quality.

There is a broad spectrum of instruments available, all of which have some implicit or explicit incentive effect, from fines to tradable permits.

This falls across a continuum ranging from very strict command approaches to decentralized approaches relying on market or legal mechanisms, as shown in Table 2.1. In these notes our focus will be concentrated on economic instruments that use pricing incentives.

2.3 PRICING INSTRUMENTS

As said before, fines or any other sanction also impose costs on polluters and users. Profit maximization would make agents equalize non-compliance and compliance costs at the margin. However, these non-compliance costs are related to mandatory standards or abatement procedures that polluters and users must follow at individual levels. Non-compliance costs will depend then on the sanction level weighted by the probability of being caught in non-compliance, that is, the expected sanction value.[2]

Sanction values are price incentives for private compliance levels and not necessarily for social cost-minimization of environmental control. The penalty level in this context must be high enough to create an incentive for all polluters and users to avoid non-compliance. Since compliance costs among polluters and users will differ, sanction value has to be set at the highest marginal cost among agents to achieve individual compliance level. Otherwise there will always be at least one agent in non-compliance status. Likewise no agent would be willing to control below the assigned individual level.

If instead all pollution and use levels were charged, each agent's level would be different. Agents with lowest costs will control more than agents with highest costs. At the end, all private agents face the same marginal cost, that is, marginal cost is equalized across agents. The total volume of control will follow the least-cost path among agents.

Horizontal equity impacts are also more interesting than with a single standard for everyone. Since the control level will increase to the point that pollution and use price equalizes marginal control cost, all controlled units will cost less than the pollution or use price, whereas pollution and use level not controlled will be charged the fixed pollution or use price.

Apart from that static result, charging all units of pollution and use will create a stronger incentive for technological change since all units will cost, not just those required to achieve compliance status.

Therefore economic instruments are in principle cost-effective, in

Table 2.1 Spectrum of policy instruments with economic incentives

CONTROL-ORIENTED →		MARKET-ORIENTED →		LITIGATION-ORIENTED
Regulations & sanctions	Charges, taxes & fees	Market creation	Final demand intervention	Liability legislation
General examples				
Standards: Government restricts nature and amount of pollution or resource use for individual polluters or resource users. Compliance is monitored and sanctions made (fines, closure, jail terms) for non-compliance.	*Effluent or user charges:* Government charges fee to individual polluters or resource users based on amount of pollution or resource use and nature of receiving medium. Fee is high enough to create incentive to reduce impacts.	*Tradable permits:* Government establishes a system of tradable pollution or resource use permits, auctions or distributes permits, and monitors compliance. Polluters or resource users trade permits at unregulated market prices.	*Performance rating:* Government supports a labelling or performance-rating programme that requires disclosure of environmental information on the final end-use product. Performance based on adoption of ISO 14000 voluntary guidelines. Eco-labels attached to 'environmentally friendly' products.	*Strict liability legislation:* The polluter or resource user by law is required to pay any damages to those affected. Damaged parties collect settlements through litigation and court system.

Specific examples of applications

• Pollution standards • Fines and other sanctions • Licensing of economic activities and land-use restrictions • Construction impact regulations for roads, pipelines, ports or communications grids • Environmental guidelines • Bans applied to materials deemed unacceptable for solid waste collection services	• Water user charges • Greening of conventional taxes • Royalties and financial compensation for natural resources exploitation • Taxes to encourage re-use or recycling of problem materials • Tipping fees on solid wastes	• Tradable use rights on resource use or pollution levels • Market-based development rights • Deposit-refund systems for solid and hazardous wastes • Tradable permits for water abstraction rights, and water and air pollution emissions	• Eco-labelling • Education regarding recycling and re-use • Disclosure legislation • Black-list of polluters	• Damages compensation • Liability on neglecting firm's managers and environmental authorities • Long-term performance bonds • 'Zero net impact' requirements for road alignments, pipelines or utility rights of way, and water crossings

Source: Seroa da Motta et al. (1999).

static and dynamic terms, in that they aim to achieve an aggregate level of environmental control without setting individual mandatory control levels. Again profit maximization will lead agents to minimize control costs. However, in this case we have a pure price incentive since polluters and users are allowed to set their own pollution and use levels. To meet the desirable environmental quality level, however, pollution prices for each unit of pollution and each source must be set so that the given volumes and locations of discharges lead to the desired aggregate ambient environmental conditions. Again, as in the case of sanctions, regulators must know agent's marginal control curves and they must know the mixing properties of the environmental medium (for example, river basin) to set the appropriate charge levels.

2.4   PRICING CRITERIA

In the case of pricing instruments such as charges, taxes and fees, apart from the flexibility issue another important feature is related to its pricing criteria. That is, what is the goal to be achieved with pricing in the chosen instrument?

Pricing criteria can be applied to accomplish three distinct goals. Firstly, achievement of the optimal use level: pricing full negative external costs in production and consumption activities to equalize prices to total full social costs. So regulators set optimal prices and the resulting pattern (volume and location) of pollution and use levels will yield the optimal degradation level. This is the so-called 'Pigovian tax' approach[3] and requires the estimation of marginal damage curves for each pollutant and use under taxation, as well as an understanding of how pollutants mix and degrade in the environment. Note that aggregate pollution and use level targets will vary by location according to the desired ambient quality in each location, and consequently charge levels will also vary. In fact, Pigovian taxes have not fully applied because of these insurmountable valuation tasks.

Secondly, improvement of cost-effectiveness: pricing pollution and use levels in order to meet a previously set aggregate level of pollution and use. Its full application requires the knowledge of marginal control cost curves (in order to set the price at the correct level to achieve the aggregate pollution goal), and it allows sources more flexibility to achieve discharge goals at lowest social costs.

Thirdly, generation of revenue: pricing natural resource uses to generate revenue. Society sets a desirable level of provision costs and prices are set to generate revenue levels that meet these cost requirements.

Although it seems obvious, it is worth emphasizing that the first and most important decision before applying an EI is to define policy objectives and their restrictions. Making policy goals compatible with pricing criteria is a crucial component of an EI that is not always recognized through its design, implementation and performance.

Moreover the reconciliation of more than one goal into a single criterion is not a trivial matter. As can be seen, each of the above pricing procedure pursues a distinct objective function and follows distinct optimization strategies, price sets and estimation complexities.

Revenues from externality and behaviour prices can be positive, but they will be bounded to the use or degradation levels rather than any budget constraint, as in the case of financing prices. This revenue would not be zero if it were to allow flexibility to users, and zero degradation or use levels were not pursued. If the prices are properly set, resultant revenue will tend to reach a stationary level that would be only altered when environmental costs and targets are revised or control costs altered.

Note that any of the criteria presented above can be set with restrictions based on distributive criteria on their objective functions, such as ability to pay and minimum free use level. That is, prices will be set with distributive weighting and allow for cross-subsidy mechanisms.[4]

2.5   COST-SAVING POTENTIAL

As said before, knowledge of cost functions is needed to set environmental prices. It is also required to estimate how much cost-saving the EI will generate. Knowing the magnitude of these benefits, regulators may persuade society and agents to cooperate, and may also evaluate the total effectiveness of the EI when the resulting implementation costs are taken into account.

It is clear that potential cost savings with economic instruments will depend on control cost heterogeneity among polluters and users, that in turn depend on location, size, technology, information and managerial skills, as well as other factors. If all agents have similar marginal control cost functions, either CAC standards or charges will lead to the social least-cost path.

For example, a simulation study (Seroa da Motta and Mendes, 1996) undertaken for the industrial sector in the State of São Paulo reveals that the use of pollution tax on organic matter, to achieve about 90 per cent of aggregate abatement level in the Tietê river basin, may reduce by 70 per cent the total abatement costs incurred by the sector when compared with a traditional command-and-control approach. Hahn (2000), presenting a survey of the experiences in market-based instruments in environmental policy in the USA reports, for example, cost savings for sulphur dioxide allowance trading in the USA of about 35 per cent.[5]

However, regulators must have in mind that these cost-saving benefits must be balanced against implementation costs including staffing and monitoring facilities. Sometimes the introduction of an economic instrument requires a great deal of institutional change in terms of expertise, tax collection and monitoring that may dissipate social control cost savings.

Although charges can first be applied based on estimated emission or use levels and next developed with self-reporting schemes, it is still necessary that managerial (inventory, sampling monitoring and so on) procedures to verify and validate reports are put in place and additional administrative capacity (bill emission, accounting and so on) is built for charge collection. This is sometimes more costly than the procedures already in place for CAC.

If CAC instruments are failing due to the weakness of institutional capacity, so will the EI if changes are not introduced. There is a temptation to reckon on EI revenues to build up the needed institutional capacity. However, revenue results, in turn, depend on institutional capacity. This vicious circle is observed in several EI experiences in Latin America, as already pointed out in Seroa da Motta et al. (1999).

2.6   ENVIRONMENTAL TARGETING AND UNCERTAINTY

Cost-effectiveness of an EI is also dependent on how close is the connection between what is priced and the quality of the environment. If the immediate source of environmental damage is not directly targeted then the policy may increase distortion elsewhere in the economy. In cases where emission or use levels are not observable, pricing will be on inputs, products or services that are related to (but not the same as) the environmental problems the policies aim to address. The so-called product or output tax for air pollution control is used in several OECD countries to tax, for example, car sales and fuel consumption (OECD, 1994, 1995), since measurement of actual individual emissions is costly. As these proxies do not target the true emission levels (for example, the car model, maintenance and driving patterns affect a car's emissions) there will be economic losses associated with mistargeting.

It is also true that the more heterogeneous the pollution and use targeted the more complex the pricing system will be, since aggregate targets may vary in spatial terms and pricing structure will need to create spatial weighting to cope with that. So homogeneous pollution from air pollution in the atmosphere is usually simpler to regulate cost-effectively than liquid effluents in water bodies.

As said before, regulators will set environmental price levels according to their assumption about agents' abatement marginal cost curves. These curves, however, are not fully observable, and also change dynamically with technological and managerial advancements within and outside the production and consumption patterns targeted by the environmental policies. In addition it must be remembered that the basis of an EI is cost differentiation and low transaction costs, so it is an instrument to be usually applied when there are quite a number of agents and consequently a great variety of costs.

So effective price setting is usually very expensive, and to keep administration costs low, experiences of environmental pricing have been applied by adjusting price levels gradually, based on results over time. That is, if regulators do not have perfect information about these curves, a trial-and-error approach may be attempted – though this risks costly errors in private sector investments (too much or too little or the wrong kinds).

Cost-effectiveness of EI depends also on symmetry of information. As in the case of CAC, if polluters and users know more than regulators about their technological abatement opportunities there will be a case for adverse selection, and consequently bad targeting in enforcement efforts.

If the application of an EI is made under uncertainty about the abatement cost of agents, it may be the case that either pollution is reduced too much, with unnecessary economic distortions, or it is reduced less than necessary and the desired environmental gains are not achieved.

Reducing pollution in excess may be politically acceptable but may, however, not occur with insufficient control. This is particularly serious when marginal damage caused by pollution or use levels is very sensitive to a small variation in pollution and use levels (that is, steeply sloped). When that is the case, high damage levels may occur due to minor price targeting errors. Therefore experiences with environmental prices are usually related to pollutants such as organic matter and suspended solids, carbon dioxide and others with such features and not, for example, in nuclear waste. How steep the damage curve is will also depend on the current assimilative capacity and carrying capacity of the environment that is to be protected.

When there is uncertainty about the cost functions and damage functions are thought to be steep, as described in the previous paragraph, it may be possible that market creation controlling quantities should be attempted instead (Weitzman, 1974; Stavins, 1996).

2.7 SUBSIDIES

Environmental subsidies in all forms are widely employed in environmental management and, depending on their form, they may also provide a marginal economic incentive in the sense that they change relative prices of environmental control by reducing its effective costs. In the short term a price on pollution and use levels or a subsidy per unit of control will lead to the same control equilibrium level. In a dynamic perspective, however, results differ since incentives for technological change are reduced because control costs are not borne by polluters and users. And new firms entering the market will be also claiming subsidies, so the subsidy approach can encourage new entry that increases pollution despite the reduction effort of the existing firms.

In equity terms moreover, a subsidy dilutes control costs among all taxpayers or reduces the costs borne by the actual users of natural resources.

In the OECD region subsidies are largely applied in the form of tax rebates as well as through credit (from the general budget or a special fund, see OECD, 1994, 1995). There are also a great number of fiscal and credit subsidies in Latin America for environmental protection on industrial investment control from end-of-pipe to R&D as well as for reforestation (Seroa da Motta et al., 1999; Acquatella, 2001).

Exemptions and tax differentiation among payees abound in the OECD (Ekins and Speck, 1999). As mentioned before, there could be equity and competitiveness reasons to protect some polluters and users with reduced environmental prices. But many of the subsidy programmes do not operate by affecting the marginal cost of compliance; they are more akin to lump sum transfers.

For example the well-known water charges (*redevances*) in France are not fully applied to agricultural use (see Chapter 3) even though this is a highly water-intensive and polluting sector. And most water charge levels in Europe are progressively set for distributive reasons. Also the famous carbon taxes applied in Scandinavian countries have exemption schemes (Ekins and Speck, 1999) for energy-intensive industries.

These subsidies are inevitable since there will always be an attempt to trade off environmental and growth targets. Regulators can behave so as to maximize political support and potential losers may have enough political power to adjust the proposed charge schedule in their favour. The phenomenon of regulatory capture is well understood (Stigler, 1971; Peltzman, 1976), and more specifically, strong evidence exists in the application of environmental CAC (see for example Gray and Deily, 1996 and Magat and Viscusi, 1990 for the USA case; Seroa da Motta, 2001 for the Brazilian case) when regulators may compromise techno-logical standards and compliance adjustment periods to accommodate equity, growth and competitiveness on regional and national levels.

Only if non-subsidized users are, in fact, overpriced to make up the subsidy amount (the cross-effect) in order to achieve either the environmental or revenue targets, does the taxpayer not end up paying the bill. However, it must be understood that cross-subsidies, even when aggregate environmental and revenue targets are met, assure equity gains only at the expense of efficiency, since overpriced agents will pay in excess of their opportunity costs.

Subsidies may also come in the form of financing facilities for payees from a fund composed of revenues of environmental charges. This is common in most water charge schemes based on river basin management (see the following three country cases). Financing of facilities can be done directly with subsidized loans for abatement expenditures or indirectly through the construction of water and sanitation utilities.

2.8   REVENUES AND INCENTIVES

Experiences within the OECD show that the use of EIs is dominated by revenue-raising aims and that they do not fully replace CACs or form part of the process towards reducing government intervention (see OECD, 1995; Svendsen, 1998; Stavins, 2002).

As previously pointed out by Seroa da Motta et al. (1999), in Latin America the use of EIs reflects a very strong revenue incentive and works with no change in CACs, almost as a complement to regulation. This does not mean that taxation for revenue will induce no change in production and use patterns. Any change in water's relative price must affect use level according to each agent's price and income elasticities.

Since compensatory measures in terms of loans and environmental investments exist, as discussed earlier it is difficult to separate the effect of financing facilities from the charge price incentive.

In Holland, where the water pollution charge level has been higher than, for example, in France, Bressers and Schuddeboom (1996) found that charge level and control level have a correlation as high as 70 per cent. That is, in the Dutch case it seems that the charge itself plays an important price incentive role.

The same was found by Ardila and Guzman (2002) for Colombia's water charge. The price incentive in the Colombian case was intentionally introduced in the newest water charge framework.

As described in Seroa da Motta et al. (2000), charges for effluent discharges and water uses (*tasas retributivas*) have been applied in Colombia since 1974 by the regional environmental agencies. The very few applications of these charges were implemented with a cost-recovery approach aimed at the operational costs of monitoring systems. The failure to expand coverage and introduce pollution and use criteria in the determination of charge levels was due to lack of appropriate design of the instrument, lack of information about impacts, lack of compatibility with the available monitoring system and lack of planning for the expansion of its coverage. Consequently the system faced vigorous public and political opposition.

No significant revenue results were accomplished. In 1993 new environmental legislation was passed in Colombia in which pollution charges are clearly specified, based on the criterion of covering full environmental costs. That is, the goal is to impose externality prices with the charge level defined according to the value of the environmental services of water and the users' environmental damage.[6]

However, this pricing criterion has never been fully implemented because of the complexity of defining charge levels, which resulted in strong legal and political opposition.

Aware of such difficulties, a decree in 1997 proposed a new charge regulation, attempting to reconcile the current legislation with a new regulation, that would allow charge levels to be determined in stages, with rates gradually estimated and implemented. Moreover attention has been paid to careful analysis of economic and social impacts for future negotiations with polluters and users. The main issues defined by this decree are the following:

1. Initially, the tax will be charged only on emissions of BOD and TSS according to a minimum tax rate.
2. Each regional environmental authority (AAR) will establish an environmental target, which will be revised every five years. A consensus-based regional process involving the different agents and communities using the resource will determine the environmental target.
3. Polluters must present an emission report every six months and random emission tests will undertake to verify, and if necessary correct, the information.
4. Tax rates will be adjusted each six months on a regional basis by a constant factor of 0.5 from 1.5 up to maximum of 5.5 until the environmental target is met.

As fully described in Rudas (2002) and Castro et al. (2001), this new legislation eliminates the cost-recovery limitations on charges that now may be fixed on an incentive basis. However, the constraints previously discussed now seem more severe. The complexity of the new way of setting charges demands a more sophisticated institutional capacity and success has been only observed in three AARs that already had good inventories and monitoring capabilities, and can count on trained staff.

Moreover, as Rudas (2002) emphasizes, in the AARs that successfully implemented the system, abatement efforts were mostly accomplished by the industrial sector that also is correctly paying the tax. However, sanitation agencies, due to their state ownership, are neither controlling nor paying. Even though judicial enforcement has been recently employed to collect tax payments due from sanitation companies, there is political opposition by the industrial users to the automatic adjustments in tax levels. They regard these adjustments as unfair since they claim that they have made the abatement effort. In sum, the cost equalizing mechanism is not understood yet.

Despite its partial implementation and the existing misunderstanding, the Colombian case is an ingenious pricing criteria that reconciles cost recovery and price incentives.

2.9   DOUBLE-DIVIDEND

In addition to the gains in cost-effectiveness in applying environmental policies, environmental taxes may be neutral by reducing the total tax burden. By reducing the tax burden elsewhere using the resulting environmental tax revenue, that is by tax recycling, it is possible to reduce general tax distortion in the economy. In sum, environmental taxation can be a less distorting way of financing public funds and diverting taxation from 'good things' (labour and capital) to 'bad things' (degradation). This opportunity is often referred to as the double-dividend.

Of course tax recycling only produces a dividend if the replaced tax is more distorting than the one taking its place.[7] Therefore the possibility of a double-dividend will depend on the existing effects of each tax under consideration and so it will vary for each country.

Such a shift is not a trivial matter, however. It depends on a solid fiscal system able to make adjustments, and also on good environmental monitoring and regulation enforcement. Therefore the implementation of appropriate economic instruments requires an institutional framework that integrates environmental and fiscal expertise and capabilities.

Such schemes that divert revenue raised to the general budget are already in place for a CO_2 tax in Scandinavian countries (see Ekins, 1999) where fiscal reforms are also in place and bureaucratic complexity is not a severe constraint.

2.10   MARKET CREATION

Instead of using prices on uses and pollution levels to achieve quantitative environmental targets, regulators could create markets where agents compete for quantitative rights (credits or permits) to pollute or use. As these rights could be tradable among agents, then an equilibrium price would emerge through trade. With perfect information and no transaction costs, both pricing and market-creation EIs would generate the same cost-effectiveness gains. Otherwise, as previously discussed, pricing is more appropriate when there is uncertainty about abatement costs but marginal damages are not seen as potentially rising rapidly with environmental degradation; whereas quantitative controls with market trading is preferred when there is more uncertainty related to the rise of damage costs.[8]

The initial allocation may either be sold at auction or distributed freely. With an appropriate auction mechanism, revenues from sales match the tax revenue that would come from prices. Since trade will guide the final allocation, the initial allocation may not be that relevant in terms of cost-effectiveness.[9] Apart from cost-effectiveness gains, regulators are also able to raise revenue and a double-dividend opportunity exists. When free distribution is followed, all income transfer takes place among agents, and consequently agents with a surplus of rights capture the rents. It is, however, controversial on equity grounds because those receiving rights above their needs make gains from trade.

Markets require also certain specific conditions for success, such as:

1. Secure property rights that enable agents to trade with no uncertainty about ownership.
2. Transaction costs of trading, information and bureaucratic behaviour must be low, otherwise trade gains are dissipated.
3. Production markets must not show distortions in terms of market power that could be passed through markets of environmental rights and also dissipate trade gains.

Industrial pollution control in the USA has traditionally followed market creation approaches since the 1970s, with a vast literature on their performance.[10] Air pollution control started with the 1974 EPA Emission Trading Programme, restricted to some air basins in the country. The programme succeeded in making agents comply with their rights, but failed to motivate trade due mainly to unsecured property rights associated with the novelty and weak legal basis of the programme. Other nationwide air pollution experiences, controlling lead, NO_x and SO_2 emissions, were later implemented with careful attention to property rights and transaction costs issues and managed to achieve quite significant trade gains (Hahn, 2000; Svendsen, 1998).

The best-known of these is the Sulphur Dioxide Allowance Trading in the US. It began as an informal market for trading SO_2 emissions rights among thermal power plants in the eastern part of the country and it has developed into a formal and sophisticated nationwide trading system among electricity plants with spot and futures markets to help identify and effect transactions (including at the Chicago Board of Trade).

Currently markets for greenhouse gases on a global scale are beginning to be built, including one within the Kyoto Protocol.[11]

Industrial air pollution control in Santiago city started in the early 1990s and has faced initially the same constraints on trade due to unsecured property rights and high administrative costs resulting from a weak institutional legal basis (O'Ryan, 2002).

Experiences with water markets in some basins in the USA had a different performance and all failed to motivate trade, mostly due to high transaction costs on the agent for transfering rights, and on the regulator's side for monitoring (Svendsen, 1998). The same diagnosis is recognized for the water market in Chile (Donoso, 2002).

In sum, market creation has a great appeal since quantitative limits on total discharges are assured (given proper enforcement) and pricing is left to trade without the need of trial-and-error approaches. On the one hand, that alleviates the regulator's demand for information on abatement costs; on the other, however, it requires a sensitive market condition and expertise and, as with direct pricing approaches, the ultimate environmental outcome may be hard to gauge in advance.

2.11   CONCLUDING REMARKS

The economic literature is prone to identify economic instruments (EIs) as a more efficient way to achieve environmental goals than quantity and technological regulations commonly called command-and-control mechanisms (CAC).[12]

The choice of an appropriate economic instrument is, however, theoretically complex and experiences with their application remain full of controversy about their effectiveness in accomplishing the proposed environmental targets. In the case of Latin America there are also serious equity issues to be considered.

In addition the most common application of EIs, including those in the OECD region, have been to raise revenue in order to fund environmental programmes and projects and/or to finance environmental management services.[13]

In Latin America and the Caribbean (LAC) the EI experiences have covered a wide range of applications that have closely followed the OECD pattern with revenue-raising aims. However, water charges have been the most advanced case and efforts are now being made to reconcile financing and incentive results.

In sum, LAC experiences have shown problems in design and implementation issues, such as:

- weak targeting and performance monitoring of environmental goals
- lack of sound pricing criteria
- poor performance in revenue collection

As previous works (Seroa da Motta et al., 1999; Acquatella, 2001) have already pointed out, fragile institutional capacity and arrangements are the main bottlenecks. The application of EIs often needs to work together with CACs and, consequently, EIs do not solve existing institutional problems. Indeed they do not reduce the burden on regulators for monitoring, and additionally they demand expertise on economic analysis and taxation, apart from requiring joint action with economic agencies.

The French river basin system has been a paradigm for Latin American experiences. This seems mostly due to the fact that the French system was created by governmental decision quite recently and implemented in a reasonable time, and was able to produce immediate results. Praising of this experience has obscured the identification of its main difficulties and constraints that, once recognized, could be of great value for followers, particularly when countries in the region are already struggling to initiate or improve the implementation of their systems.

Recent reports seem to indicate that these experiences have evolved and improved their performance. A comprehensive and careful analysis to evaluate how much this pattern has changed, as in the Colombian case by Ardila and Guzman (2002), Rudas (2002) and CAEMA (2001), is useful to indicate what has been the main driving force for such changes.

Based on that diagnosis, the following chapter will discuss the lessons of the French case that could be useful for the region's development of water EIs. Subsequent sections will analyse in detail the development of water charges in Mexico and Brazil, and their impacts on polluters' behaviour and resulting environmental control, use pattern changes, charge distributive burden, revenue path and other expected outcomes.

APPENDIX: SUGGESTED GUIDELINES FOR EI APPLICATION

When administrative costs are high and demand more institutional capacity than is available, a pricing instrument can probably face the same institutional constraints as those identified for control-oriented instruments. Not only may environmental goals be frustrated; in some cases the application of the pricing instrument results in additional budget needs rather than generating extra revenue, as expected.

Therefore much of the institutional effort on the application of the EI should be concentrated on its design in order to select viable instruments, not the 'best' or 'most desirable' ones. In doing so, regulators may adjust their existing and potential institutional capacity to the required enforcement needs.

Here will be described detailed guidelines, as first presented in Seroa da Motta (1998). The proposals are presented in such a way as may possibly be helpful to future initiatives on EI applications for environmental management in the region. The guidelines for EI formulation are presented in three phases: policy analysis phase, instrument analysis phase and instrument development phase.

a)   Policy Analysis Phase

Prior to any attempt to develop an EI, regulators may first need to analyse the policy aims and the current status of the natural resource uses.

The objective of the environmental policy
This is the most important step in formulating an EI. It is an obvious step but it is often neglected, particularly when regulators are eager to transfer a 'good' OECD experience with a particular EI to their country. Regulators must first clarify the environmental policy and the

aims for which the EI is being considered. The main outcome of this phase should be to set the policy objectives and the role of the EI, such as an externality correction and/or revenue raising. Note that an EI is by definition an instrument and cannot replace policy aims. It is designed to serve a policy and not the other way round.

Current command-and-control mechanisms
It is paramount to identify the reasons for the problems of whatever CAC is already in place to serve the environmental policy aims that the EI is supposed to replace. Very often lack of monitoring capacity, environmental and growth conflicts, as well as political constraints identified will also be barriers to EI application. In some cases such barriers may prove to be higher for price devices than for generalized CACs. Note also that ambient standards and environmental sanctions will be pertinent to EI enforcement.

Current distortionary fiscal instruments affecting the environmental goals
Sectoral policies do also apply EIs for their own aims. A subsidy or a tax on an economic activity may encourage the overuse of a certain natural resource. The removal of these distortionary fiscal instruments would theoretically be necessary to increase the efficiency of an environmental EI and sometimes may be more practical than attempting to counteract them with a new environmental EI, although sectoral political power has to be faced and may make change impossible.

Causes and sources of the environmental problem addressed by the policy aims
As was said above, economic instruments are designed to act on natural users of resources by adjusting their use levels to some desirable level or making them contribute payments to finance environmental activities. Therefore a clear identification of the causes and sources of pollution or depletion that the policy is addressing is fundamental to an understanding of users and their economic behaviour.

Environmental damage, control and opportunity cost assessments
An EI will necessarily affect environmental damage and users' control efforts (and thus marginal opportunity costs) related to the policy aims.

Therefore some estimate, even a rough one, of damages and control costs is needed before an EI is selected. Otherwise pricing devices will lack consistency.

b)   Instrument Analysis Phase

Once the policy analysis is prepared, regulators can move on to the phase of considering the selection of appropriate instruments.

Theoretical analysis

Prior to the analysis of other country or regional experiences, a theoretical analysis must identify theoretical options. Economic instruments have efficiency conditions that regulators must be aware of and must put in perspective against their own case. Although there are numerous studies proposing EIs, they often take into account these conditions in a comprehensive way. Market power, damage and control marginal cost function, asymmetric information and so on are constraints on efficiency gains that must be fully considered before the choice of an EI.

Past experiences

Regulators should review past experiences and identify the range of EIs appropriate to the policy aims. This review must consider each experience according to its relevance to the policy goals and instrumental objectives. Experiences in similar economic structure must be included, with adequate assessments of success and failure factors.

Institutional barriers

Institutional capacity for each instrument choice has to be fully assessed. Institutional analysis should take into account partnerships with other government and private organizations that the EI may affect or to which it may be of interest. Counting on possible budget reinforcements to enhance institutional capacity should also be avoided. Note that EI application may require a different expertise profile from that of the technical staff already in the institutions.

Legal barriers

The introduction of a fiscal EI may face legal barriers, not from the environmental perspective but also (possibly more difficult) in the world of tax law. Harmonization of current environmental standards and sanctions should be analysed in advance to avoid unexpected discrepancies. It is also important to avoid double taxation problems or constitutional impediments to new fiscal devices. Each EI candidate option must be analysed under such perspective.

Public perception
Some fiscal devices already have a bad reputation in people's minds because of negative past experiences or even because of lack of awareness. This is another field of analysis that cannot be neglected.

c) Instrument Development Phase

Through the foregoing analysis, regulators should be able to concentrate their efforts on a very few instrument choices, start to develop them and open public debate on them.

Monetary evaluation
An EI has to reflect opportunity costs of natural resource uses. To calculate them, regulators must follow the conventional procedures according to the kind of instrument choices. If the aim is externality correction, then it is necessary to estimate externality values. In the case of incentives, marginal control or user opportunity costs are the relevant ones whereas financing prices also require estimates of demand elasticity. Simulations and modelling exercises need to be undertaken to come up with suggested values for the chosen EIs.

Legal evaluation
In parallel with the economic evaluation of the instrument choices, regulators should also pay attention to the legal aspects of these choices. The use of EI may affect conventional property rights and consequently may require a new legal framework which can be difficult to be set up. Therefore the final choices have to be consistent with these legal aspects to avoid either a long process of legalization or future judicial disputes.

Simulation of revenue generation and distribution
Since most of the EI applications are expected to generate revenues, it is important to simulate the magnitudes of these outcomes. Note that apart from the microeconomic factors affecting revenues, such as

demand and control cost functions, revenue amounts are also dependent on macroeconomic parameters such as growth rate, exchange rate and so on. Therefore on the basis of the monetary and legal evaluation exercises, regulators should simulate revenue estimates combining micro- and macroeconomic parameters. Moreover if revenue will be distributed, as for example in terms of sectoral transfer, subsidies or loans, the scenario should also reflect these dimensions.

Economic and social impact assessment

Environmental policy is often designed to deal with a scarcity of natural resources and therefore imposes use constraints on economic agents. Very often the discussion of a proposed EI is paralysed by differing perceptions of its economic and social impacts. Although political compromise is inevitable on policy issues, regulators who are not reasonably aware of these main impacts will be trapped by articulate interest groups that seek to magnify the impacts in order to adjust the EI design or implementation for their own benefit. Consequently opportunities for efficiency and social gains can be missed. Therefore, jointly with the revenue analysis, regulators must also assess economic and social impacts and translate them whenever possible into monetary values affecting the main social and economic groups related to the policy.

Compensatory measures

Apart from this strategic behaviour, there will be some groups that lack the resources to evaluate their losses and will only become aware of them when the policy is implemented. Therefore, along with economic and social impact assessments, regulators must work out compensatory policies on distributive grounds, and growth restrictions. As was said above, although any policy instrument will create winners and losers, the use of pricing systems gives less room for the discretion of politicians and environmental agencies once it is implemented. The relation between charge costs and use levels is less sensitive to individual agreements and exemptions that are not already stated in the charging rules.

Institutional arrangements

Defining the institutional arrangements for EI choices means identifying each organization's role and commitment and the incentives for cooperation. Note that EI revenue is often a good incentive for cooperation, but secondary benefits from the successful EI application, such as public expenditure reduction and sectoral growth, may also be attractive. Regulators must find ways to confirm the capacity of each institution involved and create the necessary formal links.

Implementation planning
The EI must be planned for gradual introduction so that simulation and modelling results, as well as institutional arrangements, can be tested. National or regional policies can be implemented starting from pilot projects or experimental programmes.

Public awareness and debate
Debate with the major winners and losers from the policy and the proposed instruments should be attempted throughout the development phase to adjust decision-makers' and experts' estimates and perceptions. Fiscal devices are not well perceived by economic agents, particularly if they also restrict currently free use of natural resources. Objections on the ground of property rights often arise against any use charge; therefore public awareness of the actual costs and benefits of the policy and its proposed economic instruments has to be carefully created.

Performance indicators
Together with the implementation planning, performance indicators have to be designed to allow adjustments during the implementation process and corrections when environment and economic scenarios change. Additionally, such indicators further public awareness and acceptance.

The itemization above constitutes merely a set of suggested guidelines for EI formulation and design. Assessment of the importance of each item should be undertaken accordingly in each specific case. Of course the guidelines make demands on current institutional capacity and their full application is not always possible. Regulators can, however, use them in approaching international agencies and organizations, emphasizing the steps for which they believe technical and financial assistance is most necessary. Equally, these international institutions should orient their assistance on the same basis in order to help countries make the most of the efficiency and social gains of economic instruments on environmental management.

NOTES

1. This chapter was one of a series of papers commissioned by the Inter-American Development Bank for the Environmental Policy Dialogue and the opinions expressed in this chapter are solely those of the author and do not necessarily reflect the position of the IADB.
2. See the seminal work of Becker (1968). There are other non-compliance costs related to image effects on sales and stock values, demand restrictions and access to credit. See for example Seroa da Motta (2001).
3. Named after Arthur Cecil Pigou, economist, who first proposed it in the 1920s.
4. In any case, this would be a departure from first-best pricing.
5. See Hahn (2000) for a survey on this matter.
6. Although initially implemented for water management, these charges can be applied broadly for any environmental service.
7. The higher the marginal cost of public funding, the higher the chance of double-dividend. See for example Bovenberg and Goulder (1996), Fullerton (1997) and Parry et al. (1999).
8. This result depends on linear cost curves and correlation between these two functions is also important in choosing price or market (quantity) approaches, see Stavins (1996).
9. This is the main result of the Coase theorem, although transaction costs affect this result. We are not exploring the issue but any textbook on environmental economics discusses it in more details. Pre-existing distortions also are a source of differences between auctioned and allocated permits. Aside from the double-dividend issue, these pre-existing distortions can arise for example in the regulation of the electricity sector. If that sector has prices based on total cost recovery and permits are allocated for free, costs and therefore prices will differ from what would prevail under an auction (or full market-based electricity pricing).
10. For example, see Hahn (2000) and Svendsen (1998).
11. See, for example, a survey in IPCC (2001), Chapters 7 and 8.
12. Economic instruments here are used in the same context as market-based instruments, as they are also referred to in the literature. See for example the textbook by Baumol and Oates (1988).
13. See OECD (1994, 1995).

REFERENCES

Acquatella, J. (2001), 'CEPAL Aplicación de Instrumentos Económicos en la Gestión Ambiental en América Latina y el Caribe: Desafíos y Factores Condicionantes', *Serie Medio Ambiente y Desarrollo No. 31*, Santiago: CEPAL.

Ardila, S. and Z.C. Guzman (2002), 'Estudio Sobre El Caso Colombiano en la Aplicacíon de Instrumentos Económicos: La Tasa Retributiva', presented at the Second World Congress of Environmental Resource Economists, Monterey, CA, 24–27 June.

Baumol, W.J. and W.E. Oates (1988), *The Theory of Environmental Policy*, Cambridge: Cambridge University Press (1st edn Prentice-Hall, Inc. 1975).

Becker, G. (1968), 'Crime and punishment: an economic approach', *Journal of Political Economy*, **76**, 169–217.

Bovenberg, A.L. and L.H. Goulder (1996), 'Optimal environmental taxation in the presence of other taxes: general equilibrium analyses', *American Economic Review*, **86**, 985–1000.

Bressers, H.T. and J. Schuddeboom (1996), *A Survey of Effluent Charges and Other Economic Instruments in Dutch Environmental Policy*, Paris: OECD.

CAEMA (2001), '*Evaluación de la efectividad ambiental y eficiencia económica de la tasa por contaminación hídrica en el sector industrial colombiano*', El Centro Andino para la Economía en el Medio Ambiente, Bogotá: mimeo.

Donoso, G. (2002), '1981 Water Code: a case study, *Proceedings of the Second World Congress of Environmental and Resource Economists*, Monterey, CA, 24–27 June.

Ekins, P. (1999), 'European environmental taxes and charges: recent experience, issues and trends', *Ecological Economics*, **31** (1), 39–62.

Ekins, P. and S. Speck (1999), 'Competitiveness and exemptions of environmental in Europe', *Environmental and Resource Economics*, **13** (4), 369–95.

Fullerton, D. (1997), 'Environmental levies and distortionary taxation: comment', *American Economic Review*, **87** (1), 245–51.

Gray, W.B. and M.E. Deily (1996), 'Compliance and enforcement: air pollution regulation in the US, steel industry', *Journal of Environmental Economics and Management*, **31** (1), 96–111.

Hahn, R.W. (2000), 'The impact of economics on environmental policy', *Journal of Environmental and Economic Management*, **39** (3), 375–99.

IPCC (2001), *Climate Change 2001: Mitigation*, Contribution of Working Group III to the Third Assessment Report of the Intergovernmental Panel on Climate Change (IPCC), Cambridge: Cambridge University Press.

Magat, W. and K. Viscusi (1990), 'Effectiveness of the EPA's regulatory enforcement: the case of industrial effluent standards', *Journal of Law and Economics*, **33**, 331–60.

OECD (1994), *Managing the Environment: The Role of Economic Instruments*, Paris: OECD.

OECD (1995), *Environmental Taxes in OECD countries*, Paris: OECD.

O'Ryan, R. (2002), 'Emissions trading in Santiago: why has it not worked, but been successful?', *Proceedings of the Second World Congress of Environmental and Resource Economists,* Monterey, CA, 24–27 June.

Panayotou, T. (1993), *Green Markets*, San Francisco, CA: ICS Press.

Parry, I.W.H., R.C. Williams III and L.H. Goulder (1999), 'When can carbon abatement policies increase welfare? The fundamental role of distorted factor markets', *Journal of Environmental Economics and Management*, **37** (1), 52–84.

Peltzman, S. (1976), 'Toward a more general theory of regulation', *Journal of Law Economics*, **19** (2), 211–40.

Rietbergen-McCracken, J. and H. Abaza (eds) (2000), *Economic Instruments for Environmental Management: A Worldwide Compendium of Case Studies*, London: Earthscan/UNEP.

Rudas, G. (2002), 'Tasas para regulación ambiental: la contaminación industrial en Bogotá', in G. Rudas, *Instrumentos Económicos y Financieros para la*

Política Ambiental, Bogotá: Pontificia Universidad Javeriana, Departamento de Economía, 2002-11-06.

Seroa da Motta, R. (1998), *Application of Economic Instruments for Environmental Management in Latin America: from Theoretical to Practical Constraints*, OAS Meeting on Sustainable Development in Latin America and the Caribbean: Policies, Programs and Financing, Washington, DC, 30 October.

Seroa da Motta, R. (2000), 'Latin America region case studies in economic instruments', in J. Rietbergen-McCracken and H. Abaza (eds), *Economic Instruments for Environmental Management: A Worldwide Compendium of Case Studies*, London: Earthscan/UNEP.

Seroa da Motta, R. and F.E. Mendes (1996), 'Instrumentos econômicos na gestão ambiental: aspectos teóricos e de implementação', *Perspectivas da Economia Brasileira – 1996*, Rio de Janeiro: IPEA/DIPES.

Seroa da Motta, R., G. Rudas and J.M. Ramírez (2000), 'Water pollution taxes in Colombia', in J. Rietbergen-McCracken and H. Abaza (eds), *Economic Instruments for Environmental Management: A Worldwide Compendium of Case Studies*, London: Earthscan/UNEP.

Seroa da Motta, R., R. Huber and J. Ruintenbeek (1999), 'Market-based instruments for environmental policymaking in Latin America and the Caribbean: lessons from eleven countries', *Environment and Development Economics*, **4** (2), 177–202.

Seroa da Motta, R. (2001), *Determinants of Environmental Performance in the Brazilian Industrial Sector*, Regional Dialogue on Policy Initiatives, Washington, DC: The Inter-American Development Bank.

Stavins, R. (1996), 'Correlated uncertainty and policy instrument choice', *Journal of Environmental economics and Management*, **30** (2), 218–32.

Stavins, R. (2002), 'Experience with market-based environmental policy instruments', *Working paper 00-004*, Kennedy School of Government, Harvard University.

Stigler, G. (1971), 'The theory of economic regulation', *The Bell Journal of Economics and Management Science*, **2** (1), Spring, 1–21.

Svendsen, G.T. (1998), *Public Choice and Environmental Regulation: Tradable Permit Systems in the United States and CO_2 Taxation in Europe*, Cheltenham: Edward Elgar.

3. Country case: France[1]

Alban Thomas, José Gustavo Feres and Céline Nauges

3.1 POLICY ANALYSIS PHASE

One can view the development of an integrated and decentralized water management system in France as a response to the conflicts related to water quality degradation and the growing demand for water resources by different categories of users.

The reconstruction period following the Second World War was characterized by the acceleration of industrial development and urban growth. Environmental impacts generated by both processes were soon observed, since the regeneration capacity of water bodies was not sufficient to offset the negative effects of urban and industrial effluent emissions. As a result the water quality gradually deteriorated over the post-war period.

On the other hand, the period was also marked by increasing demands from different water users, as industrial development and urban growth proceeded. With the expansion of irrigated surface in rural areas, the same trend was observed in agricultural water demand. Moreover the acceleration of the French nuclear programme also contributed to the increase in water needs, in this case for cooling use.

In such a context, the need for a water management approach capable of reconciling the resource capacity (both in qualitative and quantitative terms) with demands for multiple uses became clear.

The legal framework was composed of measures that only acted in response to specific cases, treating user categories separately. The same differentiated approach was applied to quantity and quality aspects in water regulation. These measures established an environmental policy based on command-and-control mechanisms, some of them difficult to apply and enforce. One illustration can be found in pollution control. By law, water pollution was formally forbidden before the 1964 Water Act.

However, the strict ban on water pollution could not be put into practice because of the inevitable level of waste generated by firms and local communities. The unavoidable environmental and growth conflict condemned the regulation to lack not only effectiveness, but even basic credibility. In fact the infeasible prohibition of water pollution gave way to a regime of water pollution licences attributed by the authorities. Despite the pollution standards to be respected, in practice water pollution licences turned into water pollution rights, since it was impossible to refuse or to cancel such licences without damaging the economic development of the region. A common case would be a firm holding a water pollution licence being prosecuted for environmental damage, in spite of the fact that this same firm would not have violated the emission standards stipulated by the administration.

In addition to the low effectiveness of the regulatory mechanism, the legal framework could not take into account the issue of negative externalities. Polluters were not given incentives to exert the necessary financial effort in order to treat wastewater, which would benefit downstream users. In order to satisfy their growing demands for potable water, downstream cities and towns were obliged to build costly water treatment facilities.

In short, the regulation was not adapted to the context of growing conflicts related to water availability and water quality, and an integrated management system was called for.

The Water Act of December 1964 deeply modified the French water management system. The new approach to water management policy established by the law was based on two general principles: decentralization and planning. Decentralization was based on the idea that water management organization should reflect the physical unity of water bodies in order to account for potential sources of conflicts. To handle the externality problems linked to water pollution and conflicts of use in an integrated approach, the river basin was chosen as the basic administrative unit. Planning was to provide consistent decisions at the river basin level and to introduce a medium-term perspective on water management.

The decentralization principle is put into practice by the creation of Water Agencies (WA) and River Basin Committees (RBCs) in each of the six French river basins (see Table 3.1 for their corresponding surface area). While the agencies are intended to perform executive functions, RBCs act as consultative bodies and are sometimes described as local 'Water Parliaments'.

Table 3.1   French river basins

River basin	Surface area
Adour-Garonne	115000 km^2
Artois-Picardie	19562 km^2
Loire-Bretagne	155000 km^2
Rhin-Meuse	31500 km^2
Rhône-Méditerranée-Corse	130000 km^2
Seine-Normandie	100000 km^2

Source: *Guide de l'Eau* (2001).

The RBC is responsible for analysing any subject considered relevant to the river basin. The various agents concerned with water management are represented: communities, water users and the central administration. The principle of decentralization is clearly observed in its composition, since the central administration holds less than half of the representatives. The committee constitutes the locus where conflicts between stakeholders related to water quality and water availability can be solved. By assembling the interested parties established in the river basin, the decisions of the RBC are expected to reflect the general interest of all users and stakeholders.

Water Agencies are the executive branch of the RBCs. Financially autonomous, they are in charge of collecting water charges and, through loans and subsidies, financing private and municipal investment projects intended to reduce pollution and increase water availability. Financial assistance is aimed at providing polluters (firms and communities) with incentives to undertake pollution-reducing and water-saving investments. Eligible investments must be in accordance with the priorities defined by the WA (and later approved by the RBC) in the multi-year working plan, established for a five-year period. The plans are supposed to reconcile demands for the multiple water uses and identify the most important pollution reduction actions and water availability measures to be taken in the river basin, providing guidelines for the WA intervention during the period.

The law also innovated by introducing water charges. These charges have two objectives. Firstly, to finance the investments defined in the intervention programme established by the Water Agency, and secondly, to induce water use efficiency. The water charges are suggested by the WAs and must be approved by the RBCs. This approval is supposed to guarantee that the interested parties accept these charges, hence legitimating their use as economic instruments.

The water charges are collected on the basis both of water withdrawals and of effluent emissions. In order to calculate the emissions charge, industrial and domestic effluents are systematically measured (or estimated) and taxed according to the effluent emission rates defined in the five-year programme. The withdrawal charge is based on the user's quantity of intake water. Water charges were supposed to act as an instrument to promote use efficiency and not merely as a pure revenue-raising instrument. However, as it will be discussed later, until recently revenue-raising purposes have prevailed over use-efficiency ones.

The revenue collected is then used to finance water-saving and pollution abatement investments undertaken by water users, according to the directives of the working plan defined by the WA.[2] The financing mechanism can assume several forms, such as subsidies or loans with preferential interest rates, among others. This financial support also has an incentive component, as it reduces the cost of investing in water pollution abatement facilities that could not have taken place without this financial scheme.

The transformation of French institutions, especially regarding the Decentralization Act of 1982, and the need to transpose the European Community directives into the French legal texts, pointed to the need to update the Water Act of December 1964. This was accomplished in the Water Act of January 1992, which should not be interpreted as a structural reform of the preceding Act of 1964, but as its complement and in continuity with its principles. The institutional structure of the water management system based on the river basins and their respective Water Agencies and Committees was not modified. The Water Act was aimed at establishing a more balanced water resource management system, giving more nearly equal weight to environmental and economic interests. More precisely, the concept of well-balanced management seeks the preservation of water ecosystems and wetlands, the qualitative and quantitative protection of water resources, and the attainment of multiple water uses. The 1992 Water Act also stressed the need of effectively implementing the 'polluter pays' and 'user pays' principles through water charges designed to make the agents internalize water pollution and use costs.

The objectives proposed by the integrated management approach are the result of the transposition to the French law of the European Community directives. Compliance with these European directives has been orientating French water policy toward the strengthening of the incentive properties of the water charge system by putting into effect the 'polluter pays' principle. Enforcement of the 'polluter pays' principle is intended to be achieved through (1) more control on individual emissions and abatement practices, (2) the enlargement of the set of pollutants to be taxed, and (3) the handling of agricultural water pollution.

3.2   INSTRUMENT DESIGN PHASE

3.2.1   Theoretical Basis

It is well accepted that a price on pollution and natural resource uses will induce polluters and users to adjust pollution and use levels to their individual least-cost paths. Implementation of such charge may nevertheless be confronted with common problems of asymmetric information on pollution level and possibly abatement cost, as well as of political acceptability. In France this is mainly due to the system of subsidies complementing the charge application. This system of course reduces the impact of the charge and it has been said that it is not the charge that provides the incentive for adaptation, but rather the subsidy.

3.2.2   The Design Framework

Regarding the design of the water charge as a policy instrument, several important points must be made. First, WAs design different water charges, both for withdrawal and for effluent emissions, depending on the category of users (farmers, households or industrialists). Second, the revenue collection is also different and a distinction has to be made between residential users on the one hand, and farmers and industrial users on the other. Industrialists and farmers are directly charged by WAs through two different channels, one for water withdrawal and one for pollution. The majority of farmers are not subject to charges, for reasons that will be later developed.

Residential users are charged by the municipal water utility through a pricing system which, since the 1992 Water Act, has to be based on actual water consumption alone, as measured in cubic metres. The price of water is the sum of the price charged for the service of distribution, the price of sewage treatment, and other charges that include the charge on water use and effluent emissions collected by the WA. The effluent emissions charge is to cover the cost of pollution generated by residential uses and the use charge is based on the quantity consumed. Both charges are designed by the local WA.

These latter charges are treated as components of the price of water and residential users do not pay the WA directly but through the local supplier, who acts as a tax collector for the WA. As a result, water and sanitation municipal services commonly increase charges to cover the water cost.

The effluent emissions charge

For industrialists, the effluent emission charge is proportional to the number of pollutant concentration units (in general, kilos per day). When actual pollution is not observed, an estimate is used instead, based on the firm's production data. To this end, the WA makes use of an industry mapping between production activities and estimated average pollution level for different pollutants (biochemical oxygen demand, suspended solids, nitrogen and so on). In the case of residential users, the charge is a fixed amount to be paid for each cubic metre consumed. Broadly speaking, it is based on the ratio of the expected total amount of pollution generated by a local community to its expected total water consumption. The formula also includes two types of coefficients: a coefficient for each pollutant depending on its potential damage for the environment, and a coefficient that varies depending on the total permanent and seasonal population.

In principle all polluters are liable to the charge in order to guarantee some form of equity. However, setting aside the farmers, it is still the case that some public services (local administrations, public facilities, public parks, fire brigades and so on) are exempted from the charge.

Some prior information is needed in order to design the charge correctly. However, this prior information required for designing the charge rate is both difficult and costly to collect. The necessary information includes:

- Identification of all substances that could damage the environment, that is, all pollutants that will be targeted by the charge.

- The cost of abatement associated with each type of pollutant identified in the first step has to be estimated. This cost is quite technically difficult to determine and the ones who know it (the polluters) are often hiding the information.
- The preferences of the consumers (present and future generations) and their valuation of the environment have to be known in order to make it possible to measure welfare effects. This is of course very difficult to determine and significant heterogeneity among consumer preferences across different regions is to be expected.
- A charge should be based on observable elements and would call for the implementation of a system of costly continuous measures of effluent emissions.

Ideally charge rates for pollutants should be based on scientifically measured reports. There is no consensus at this time on the precise impact of each pollutant; therefore coefficients associated with each pollutant differ from one Agency to another and it is also the case that some pollutants are charged by one Agency and not by another.

A common structure of pricing should be adopted in order to guarantee equity in terms of pricing principles. Scientific and technical analysis should also be encouraged to reach a consensus at the national level on the relative risk associated with each pollutant.

If the charge was designed to match the marginal cost of abatement, we should observe a wider dispersion of the charges between local communities and Water Agencies. As remarked by Neira (1995), this is not the case in France, where we observe that the rate per cubic metre follows a common tendency in all basins.

The use charge
The fundamental aim of the use charge is to cover the cost of resource management including financing of storage dams, supply networks and irrigation equipment. A distinction is made between surface water and groundwater. For surface water, the charge is the sum of a charge based on the pumped volume and a charge based on net water consumption. Each of these two elements is multiplied by a constant rate and a rate depending on the type of use. For groundwater, there is no distinction between pumped and consumed water.

To provide adequate resource allocation, actual consumption should be observed for individual users, but accurate data on net consumption are sometimes difficult to obtain, especially for farmers and local communities. This is why estimated use levels are computed from production or population data and are used to calculate water charges rather than actual consumption levels. Large discrepancies between charges per unit of use among river basins are observed. It can be the case that both surface water and groundwater are charged, or that only one type of source is liable to the charge. The coefficient applied to each of these two sources can also vary significantly in time and from one WA to another.

Different rates are applied depending on the use, and this distinction is not justified. For instance it may benefit farmers but penalize residential users and industrialists. The difference can go up to a factor of 40 for the same volume of water withdrawn from the same source.

Table 3.2 presents unit effluent emission and water use charges for the six WAs, in 1992. Note that unit water use charges may vary within the WA, according to the time of year.

Table 3.2   Unit effluent emission and use charges in 1992

Water Agency	Suspended solids	BOD	Nitrogen	Phosphorus	Water use
Adour-Garonne	158 30	254 96	226 27	106 76	[0.12; 0.18]
Artois-Picardie	126 00	252 00	143 00	675 00	[0.10; 0.31]
Loire-Bretagne	92 11	141 70	173 00	272 54	[0.16; 0.36]
Rhin-Meuse	103 19	206 37	141 59	235 53	[0.15; 0.30]
Rhône-Méd.-Corse	80 00	240 00	120	300 00	[0.05; 0.30]
Seine-Normandie	113 93	249 69	213 69	NA	[0.09; 0.26]

Notes: In French francs per kilo-day for suspended solids, BOD, nitrogen and phosphorus, in French francs per cubic metre for water use. BOD: biological oxygen demand. NA: not applicable.

Source: *Guide de l'Eau* (2001).

3.3   INSTITUTIONAL BARRIERS

Potential barriers to the design of an efficient water charge system can be related to institutional issues, namely the composition of RBCs, and the level of knowledge available to WAs concerning water environment, external costs and consumers' behaviour.

The composition of the RBCs plays an influential role in the determination of general principles adopted in the definition of water charge levels. In fact, it is tempting to transfer a part of the financial charge to the actors that are under-represented, or even not represented at all, in the institutional apparatus. This point finds a good illustration in the financing of the agricultural pollution programme (PMPOA) whose financial costs are supported by the final users.

The second point concerns knowledge build-up by WAs. Generally speaking, and contrary to the objectives established by the Water Act of December 1964, Water Agencies' knowledge acquisition has progressed poorly. This knowledge was supposed to improve the economic foundations of water charges. It was intended to be developed on water environment, external costs, and the behaviour of polluters and users. None of these research directions were supported by stable and significant funding. Amigues et al. (1994), in a report for the Commissariat Général du Plan, state that 'if nowadays it seems difficult to know precisely, or even approximately, the effects of a treatment facility on the water environment, this difficulty can be attributed to the low funding levels assigned to modeling these effects'. The report written by Martin (1996) on sustainable management of groundwater in France also points out the insufficient funding levels allocated to research. According to the report, this fact, given the increasing groundwater intake, especially for irrigation purposes, prevents implementation of a sustainable groundwater management policy.

The same failure to increase knowledge can be seen in the study of external costs. Although there exist numerous methods to estimate environmental damage, their application can still be considered rare in France. These methods, even if they remain controversial, could have their use intensified in order to support decision-making and promote the rationalization of the water charge system currently in place.

Finally, the WAs have not given enough attention to acquiring knowledge of water consumers' behaviour. Such lack of knowledge limits the development of a water charge system with adequate incentive properties. It could be argued that, since WAs are to some extent subordinated to RBCs, user representatives within committees could have pressed the Agencies to increase research effort. However, although WAs have specific budget lines devoted to research funding, the vast majority of funded projects still concern technological improvements of wastewater treatment (biophysical and biochemical processes and so on).

3.4   LEGAL BARRIERS

Since their introduction, the constitutionality of water charges has been debated. The dubious juridical nature of water charges has raised some problematic issues. It has prevented, for example, the application of necessary or envisaged modifications in water charges.

In 1991, three WAs (Adour-Garonne, Rhône-Méditerranée-Corse, Rhin-Meuse) tried to introduce a specific charge for extractive industries (sand, gravel) on the ground that the activity of these industries modified the regime of river streams. The concerned firms then referred to the Conseil d'Etat (French Supreme Court), which cancelled all measures taken by the WAs to implement the charge. WAs were forced to reimburse all revenue collected previously from this particular charge. The Conseil d'Etat understood that this specific category of extractive activity was not to be subject to the additional water charge, not because its activities have no impact on water resources, but rather because the corresponding legislation was still deficient and incomplete.

3.5   INSTRUMENT IMPLEMENTATION PHASE

3.5.1   Implementation Path

The implementation path of the policy has been strongly conditioned by the way perception of water charges by environmental managers has evolved through time (see the institutional factors in section 2.3). Implementation may be analysed in three periods.

During the first of these, between 1967 and 1992, the water charge system was progressively accepted by local communities and industries regarding the level of unit charges, set of chargeable pollutants and the charge collection system. Of course this is mainly due to the original vision of the water charge as a moderate financial contribution, which could somehow partly be returned to polluters by means of abatement cost subsidies and the fact that mutual consent on the level of water charges by all local stakeholders is favoured (through the RBCs).

The second phase, starting in 1992 and contemporaneous to the French Water Act, is characterized by a global need to correct for major discrepancies between environmental objectives (at a national but also European level) and actual achievements of the Water Agency policy. Although the WA system has not been questioned, public policy-makers have recognized a need for more control on individual emission (and abatement) practices and for a redefinition of the set of pollutants to be charged. It is interesting to note that the 'polluter pays' principle is now openly mentioned as the underlying principle of the system, meaning that the original vision of the WA as a 'union of water users' is to be progressively abandoned.

The final phase, starting in the late 1990s, builds around the 2002 project for a new Water Act. This project was aiming at accounting for socio-economic consequences to households of water price increases and, more importantly, at incorporating a larger part of the agricultural sector in the water charge system. The restructuring of the agricultural sector has had major impacts on overall nitrate contamination of groundwater and surface waters, but agriculture had only been partly subjected to charges. Since the beginning of the 1990s, only some specific breeding activities had to pay emission charges (pigs, chickens). For cereal and beef cattle producers, a project initiated by French MPs in the late 1990s intended to charge directly actual contamination of water resources by nitrates and pesticides. Because of the obvious problem of measuring non-point source pollution, the project was modified to allow farmers to contribute by means of a tax proportional to production and land occupation patterns. This project was, however, abandoned by the new political majority after the general elections in the spring of 2002. Government officials claim that a new project will be proposed in 2003 containing a redefinition of charges including farmers in the system, incorporating a simplification of the water charge rates and stressing the need for a decentralized system.

In sum, the Water Agency system is improving in monitoring and control, imposing higher charges but also offering more subsidies.

The move toward a 'more efficient' environmental policy by WAs at a local level has been taken in a period where the industrial and residential structures of the country have experienced significant changes. With the decline of local production capacity in areas where traditional industries were the majority (textile in the north and the south-west, steel mills and heavy chemistry in the east, and coal extraction in both), the picture of industry has changed over the past 15 years. In particular the pollution picture has been characterized by a lower per-plant emission rate for major and well-known pollutants. The overall industrial growth rate has been about 2 per cent a year since the beginning of the 1980s, and during the same period, suspended solids have decreased by 4.3 per cent a year, and BOD (biochemical oxygen demand) by 3.6 per cent a year. Nevertheless at the same time a new generation of pollutants is becoming increasingly important (absorbable organic halogen, heavy metals). In addition the need for enhanced energy production has led to an increase in temperature in many rivers because of the way cooling needs have been met at fossil and nuclear power plants.

Turning now to water use charges, industries have since the beginning been incorporated in the water charge system, but with a charge representing a very limited fraction of their total water bill. The most interesting trend, however, is for farmers, who are now required to equip their farms with metering devices, with exceptions for small farms. The level of the charge for irrigation water use is typically very moderate and is of a multi-block nature. The fact that unit charges are bound to increase has been recognized for many years now, but it is expected to have a limited impact on that population of farmers, who will be concerned with such a system for two reasons: first, because of agricultural subsidies for irrigation, and second, because of the setting of a relatively 'high' minimum perception level of 7000 cubic metres a year.

3.5.2   Environmental Outcomes

Recent trends in water quantity and quality can be illustrated by the following figures. Environmentally sensitive zones now encompass about 50 per cent of the whole territory, compared to figures around 20 per cent registered during the 1990s. For one of the largest groundwater resources, in Alsace, more than half of measurement stations indicate nitrate concentration higher than the target value of 25 mg per litre, and 12 per cent have a nitrate concentration of more than the regulatory (and mandatory) limit of 50 mg per litre. In the south-west (a major area for grain production), most water extraction outlets have maximum observed nitrate concentrations greater than 50 mg/l.

About 90 per cent of French rivers and streams are contaminated to some extent by pesticides (10 per cent of these streams have pesticide concentrations greater than the accepted health standard, as do 40 per cent of groundwater extraction outlets). A yearly average of 646 000 tons of nitrogen are carried by major rivers to the sea, which is of concern given the international commitments of the country toward its European neighbours. Finally, about 5.6 billion cubic metres a year are used for irrigation (12 per cent of total extracted volumes), of which 88 per cent is from surface waters. Net consumption from agriculture represents 43 per cent of the total.

While agricultural-related contamination has increased, positive environmental outcomes of water charges in France have been significant for industrial sources in regions where water quality has been problematic in recent decades. In all river basins, the upward trend in industrial pollutant concentrations has been reversed, although difficulties remain in regions where monitoring of abatement activities and emission self-reporting schemes have not been successfully applied.

It is difficult to provide an assessment of the actual contribution of charge schemes to this limitation in overall pollution levels because of the interaction with the command-and-control regulatory policy at national (and since the 1980s regional) level. Nevertheless one can adopt the view that, although command-and-control policy has been the major determinant in abatement control in the industrial sector and domestic sewage treatment, the emission charge system can be seen as the most important factor for firms, after abatement equipment is in place. Recent empirical studies (for example, Lavergne and Thomas, 2004) provide evidence for the significant impact of effluent emission rates on abatement equipment operation by industries.

Even if the charge scheme is shown to have had an impact on emission reduction, the relationship with actual environmental conditions is not straightforward. A reduction on effluent industrial emissions can only improve water quality if other pollution sources remain constant. It is not relevant to analyse the impact of changes in water charges by overlooking the way pollution is defined at an individual level. Charge payment based on average emission levels by industry is becoming less predominant and Water Agencies now seek to have firms (and local communities) internalizing the cost of emission measurement (self-reporting policy).

In sum, the observed trends in water quality are revealing. On the one hand, surface waters that are mostly affected by industrial and residential pollutants have experienced a significant reduction in their pollution levels. On the other, groundwater, which is more sensitive to agricultural emissions, has not.

Nitrate concentrations from agricultural sources have increased or remained stable, both for surface and groundwater, even in cases where agricultural land has decreased. Agricultural pattern changes are responsible for that. Land areas under pastures and fodder crops have decreased by 4.3 million hectares since the 1970s, whereas areas for cereal and oilseed crops have increased by 1.9 million hectares. At the same time, irrigated land has increased by 47 per cent between 1988 and 2000, led by grain production. Hence a major result of the restructuring of the French agriculture sector has been the conversion of significant land surface to higher nitrate-using and water-intensive crops. Of course predicting environmental outcomes regarding water quality would be irrelevant for agriculture because, as pointed out above, the majority of farmers are not incorporated in the water charge system.

Unit water use charges have also been too low to induce a significant reduction in agricultural (irrigation) use patterns. In contrast, water charges are the major component of water input expenditures in the industrial sector.

The Ministry of Environment in 1993 collected data on investment decisions made by local municipalities. These showed that environment preservation is one of the priorities of the surveyed municipalities, and that two-thirds of them put sewage and water treatment as the highest priorities. Financial assistance by the Water Agency is described as the third factor (after the lack of action in the past and the regulation via emission standards) that induced them to put sewage and treatment of water as a priority. If subsidies globally represent 52 per cent of total investment in the municipalities surveyed, it is clear that priority is given by Water Agencies to pollution abatement: 68 per cent of the total cost of investment directed toward pollution abatement is subsidized by Water Agencies' grants. As far as water resource management is concerned, the Ministry of the Environment (2001) reports that only 34 per cent of total investment cost is subsidized by the Water Agencies. Modernization of the supply network appears less important for local communities. It must be noted that the 1991 European directive on urban wastewater (see European Community Council (1991)) has also imposed the building of wastewater treatment plants in all local communities (except for very small ones). The main constraint to investment, however, has been community inability to cope with high costs of sanitation and water projects.

3.5.3   Economic and Social Impacts

Agriculture and industry

In most parts of the country, water is not scarce enough to justify significant levels of water charges. Furthermore many industrialists are able to use private wells instead of community distribution networks, and as a consequence the cost of water expenditures is typically very low compared to other inputs (labour, energy, materials and so on). The impact of recent (since the beginning of the 1990s) increases in water use charges on firms' financial situation has therefore been negligible. On the other hand, the significant increase in effluent emissions charges is likely to have had more important consequences for firms that have not invested in water abatement facilities.

The water charge system by which polluters can be partly compensated and funded from revenues of water charges when investing in abatement facilities is a major advantage, in the view of polluters, compared to a standard command-and-control rule. Nevertheless with such subsidies the efficiency of the whole system is to be questioned, in terms of the opportunity cost of public funds.

As noted above, farmers are supposed to be charged for pollution generated by nitrogen and pesticide use, on a proportional basis related to acreage and/or production. When or if such a policy is implemented in the near future, it is expected to have a very significant impact on farmers' financial outlook. Recent simulation studies can be invoked to provide evidence for the fact that pig and beef cattle farmers will suffer the most from such nitrogen taxes. Note that even if the ultimate goal of such charges is to promote higher water quality levels in areas with significant non-point source pollution from agricultural sources, this policy instrument is not typically a water charge, as it is based on agricultural inputs other than water use (such as chemical nitrogen fertilizer).

Residential users

As explained before, water charges are included in the water bill. Table 3.3 presents the detailed breakdown of the average residential water bill per year in France (for an average annual consumption of 120 cubic metres) for the period 1995–2000.

Table 3.3   Breakdown of the average residential water bill

	1995	1996	1997	1998	1999	2000
Water supply	120.74	125.31	128.36	129.73	131.41	133.55
Use charge	4.66	4.94	5.10	5.12	5.08	5.35
Treatment cost	84.61	89.18	93.60	96.20	98.02	99.40
Effluent emission charge	38.59	43.31	44.32	46.34	48.03	49.27
National and local taxes (VAT, etc.)	25.46	28.51	29.58	29.73	29.88	30.03
Total	274.06	291.25	300.96	307.12	312.42	317.60

Notes: Prices are in euros. Figures for an average consumption of 120 m³/year. Water supply and treatment costs are paid to the water utility operator, while use charge and effluent emission charges are paid to the Water Agency.

Source: DGCCRF (2001).

Between 1995 and 2000, the total amount of payments by residential users to the WAs (use and effluent emission charges) increased by 26 per cent while the other elements of the water bill increased by 16 per cent. The share of each element remained almost constant in the period. The charge for effluent emissions was around 16 per cent of total water bill in 2000 while the use charge represented less than 2 per cent of the total bill. Priority has clearly been given to pollution abatement.

Measurement of impacts on people's income due to water charges is not available. However, some case studies (Nauges and Reynaud, 2001; Nauges and Thomas, 2000) have estimated the price elasticity of water demand and change in welfare following a water price variation. Using two samples of French local communities,[3] it has been found that French residential consumers are sensitive to water price. Water demand is found to be responsive to its own price, even if the estimated price-elasticity values are quite moderate (between –0.08 and –0.22, which means that if the price of water increases by 10 per cent we expect a decrease in consumption by 0.8 to 2.2 per cent).

From these estimates it was then possible to calculate change in welfare due to a change in water price. The change in welfare is measured by the compensating variation, defined as the amount of money that would compensate the consumer for the change in price, leaving all other things equal. Welfare losses are measured as money values in terms of a proportion of a consumer's water bill. It is estimated that a 10 per cent increase in price would lead in both samples to a welfare loss equivalent to 7 per cent of the total water bill whereas a 20 per cent increase would imply a welfare change representing 13 to 14 per cent of total bill. Therefore it can be expected that welfare impacts of water charges follow the same proportion since they are similar water bill components. However, it must be noted that the above-mentioned compensating variation measures only indicate losses of welfare assuming everything else remains constant, and therefore they do not take into account welfare gains derived from environmental quality improvements.

It should be noted that the Water Act project of 2002 explicitly accounted for socio-economic consequences of the significant increase in the price of residential water for low-income households. A provision was made to exempt this category of consumers from water charges. The local communities would then have been compensated for these unpaid liabilities by a mutual fund for 'water solidarity' partly financed by the government. Whether such provision will be preserved in a new Water Act in preparation remains uncertain.

3.5.4 Institutional and Legal Factors

There are many important institutional and legal aspects to the success or failure of the charge system, during the implementation phase. In fact, agricultural pollution was supposed to be integrated in the water charge system, but it was not because of institutional constraints. Although the Ministry of the Environment could in principle enforce environmental regulation on farmers, such as collecting water taxes or implementing a system of quotas on nitrogen fertilizer and pesticide use, expected compensation in the form of subsidies had to be considered by the Ministry of Agriculture. However, these subsidies could come into conflict with European agricultural subsidy policies already in place. The whole question was then to decide whether agricultural revenue and price support policies had to be coupled with environmental objectives at the national level.

The fact that a market for water rights has never been implemented is because water users do not own the right to sell water in France; they only possess the *usus and fructus* properties, inherited from Roman law. Therefore unless a major change in French law is decided upon, the setting up of a market for water rights is not to be expected.

The decentralized organization of River Basin Committees and Water Agencies, including representatives of the sector, has helped much in the acceptance of the water charge system by the industrial sector. Nevertheless, as pointed out by CGGREF (1994), the likelihood of collusion between Water Agencies and industrialists is seen as a weakness of the decentralized approach.

In practice, monitoring and control activities for law enforcement purposes, including inspection of industrial plants and control for compliance with emission standards, is not under the charge of Water Agencies. Environmental regulation is under the supervision of national or regional environmental agencies, completely independently from the Water Agencies system. This separation between local water institutions with river basin attributions (the Water Agencies) and parallel governmental institutions for law enforcement is definitely a drawback. Usually useful information is not conveyed from one agency to another and policy orientations may differ on the same subject. This does not mean that environmental objectives from both authorities, translated into environmental monitoring on the one hand, and water charges on the other, should in most cases be contradictory. Rather, water charges are often considered a means of reaching an even higher objective of water quality than would have been obtained with environmental monitoring alone. Although both policies seem to be complementary in practice, the cost of estimating effluent emissions by either party could certainly be reduced if information at the plant level was shared between the Water Agency and the law enforcement authority.

3.5.5 Public Awareness and Debate, Political Acceptability

The public has gradually become fully aware of the fact that although water is not a scarce resource *per se*, proper quality and availability standards require higher prices than those prevailing in the past. The important increase in water bills for residential users has been generally understood as the consequence of the implementation of European directives on urban wastewaters and human water consumption, which require important investments in abatement and raw water treatment outlets in local communities. The debate has been mainly on the subject of 'who was responsible' for these price increases. Private companies in charge of water utilities claim that this increase is mainly due to wastewater treatment, not to water production and delivery to customers, and that charges have increased more than their own share in the total water bill.

A second point of debate has been the role of non-point source pollution from agriculture. Although the 1992 Water Act stipulates that quality water should be made available to all customers without restrictions, large areas in the country, especially in the Northwest (Brittany) still face very poor water quality standards because of intensive animal breeding (pigs and poultry) in the region. The contradiction between a higher price for water and the very poor quality of tap water has led populations in small towns to refuse to pay their water bills, and to ask for refund of their bottled water expenditures.

Recent public reports by the 'Cour des Comptes' and the 'Commissariat Général du Plan' in 1997 revealed that although the French water management system was presented abroad as a very efficient one, 'it is not organized to reach qualitative and quantitative objectives as defined by the government, at the least possible cost'. As WAs are decentralized, they are not concerned with government cuts in the general budget, and water charges increased by 22.7 per cent a year between 1988 and 1994. Such an increase left many significant inequalities and inefficiencies unresolved: for example animal breeders are considered to generate a pollution level equivalent to 254 million residential users, whereas they only contribute 2 per cent to the funding of the water policy. As a consequence residential users are propor-tionally taxed much higher for effluent emissions than farmers and even industrialists.

Concern and organized protest from residential consumers is growing mainly because of equity issues. Residential users are the main contributors to the Agency budget, while being at the same time the less-favoured stakeholders in the system. This situation is more and more often criticized now as the price of water has been increasing sharply since 1992. In Tables 3.4 and 3.5, one can verify that, during the period 1992–2001, residential users accounted for more than 80 per cent of the total water charge revenue collected by Water Agencies. Their contribution to WA revenue has even increased in the VII Working Plan, contrary to industrial users whose relative participation has decreased.

Figures for agricultural users contrast sharply with those of residential users. Agricultural users provide a marginal contribution of around 1 per cent of total water charge revenue, in spite of the fact that farmers' consumption represents 70 per cent of total net water

*Table 3.4   Revenues from water charges collected by Water Agencies,
VI Working Plan 1992–96 (in million French francs)*

	Water pollution charge	Use charge	Total	User share (%)
Residential	24899	4746	29645	81.1
Industry	4998	1686	6684	18.3
Agriculture	0	245	245	0.6
Total	29897	6677	36574	

Source: Ministry of the Environment (2001).

*Table 3.5   Revenues from water charges collected by Water Agencies,
VII Working Plan 1997–2001 (in million French francs)*

	Water pollution charge	Use charge	Total	User share (%)
Residential	35614	6361	41975	83.7
Industry	5437	1910	7347	14.7
Agriculture	554	269	823	1.6
Total	41605	8540	50145	

Source: Ministry of the Environment (2001).

consumption in France. On the other hand, they are the main
beneficiaries of Water Agency transfers, as it is indicated by the ratio of
transfers over charges for the period covered by the sixth working
programme of the Water Agencies, shown in Table 3.6. The discrepancy
between agriculture and the two other sectors is clear. As water charges
have been rising significantly since 1992 for residential consumers, the
privileged treatment given to the agricultural sector is now being
seriously questioned.

3.5.6   Revenue Generation

The definition of the total water charge revenue is conditioned by the
five-year working plan established by the Water Agencies. The general

Table 3.6   Subsidies and charges for the VI Working Plan 1992–96 (in million French francs)

Users	Charges	Subsidies	Ratio subsidies/charges (%)
Residential	29 645	35 232	119
Industry	6684	7317	109
Agriculture	245	1076	439
Total	36 574	43 625	

Source: Ministry of the Environment (2001).

mechanism is characterized by the Decree No. 66-700 of September 1966: 'The total amount of water charges … is to be fixed in function of the expenditures of all nature incurred by the Water Agencies, in the framework of their intervention programs.'

Figure 3.1 illustrates how the integration between working plans, investment needs and water charges works.

As a consequence of this 'expenditure-driven' principle, one can observe in Table 3.7 that the revenue outcome of water charges follows closely the total budget of the working plans. Subsidies and revenues increased steadily during the period 1977–91 without causing significant opposition from water users, since the charges represented only a moderate financial contribution. The VI Working Plan, covering the period 1992–96, is characterized by a high increase in the subsidies by the WAs, as a result of the implementation of European directives on urban wastewaters and human water consumption. The subsidies and consequently the charges almost doubled during the period. The expenditure share in the total water bill due to water charges paid to the WAs increased from 7 per cent to 14 per cent.

The relative weight of the financial burden due to water charges is gaining in importance, and so do complaints by residential water users, as pointed out in the previous section. Faced with growing complaints, the Prime Minister decided to interfere in the design of the VII Working Plan, determining the adoption of a revenue-driven mechanism in order to control for water charges increases. That is, first the revenue to be attained is determined and then the intervention programme is defined. This revenue-driven mechanism, at the cost of a one-year delay in the

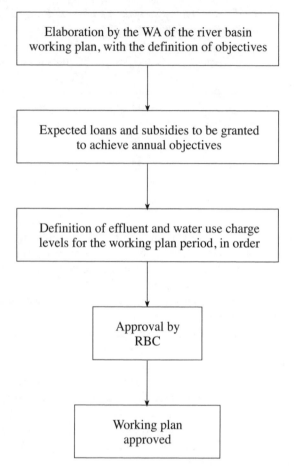

Figure 3.1   Interaction between working plans, investment needs and water charges

compliance to European water and sanitation directives, is supposed to facilitate acceptance of water charges by public opinion.

One important normative point to be noted concerns the issue of the destination of funds collected by Water Agencies. In the late 1990s, a governmental project was drafted to let funds collected by Water Agencies be diverted to the French Ministry of Finance, as part of the global public budget. The opposition was fierce and the precise nature of the water pollution charges was at the heart of the debate. For

*Table 3.7   Evolution of charges collected and subsidies paid by Water
Agencies, 1977–2001 (in billion French francs)*

	III Working Plan 1977–81	IV Working Plan 1982–86	V Working Plan 1987–91	VI Working Plan 1992–96	VII Working Plan 1997–2001
Subsidies	14.3	16.3	22.3	40.7	57.0
Charges	14.3	14.6	21.1	40.1	50.9

Source: Ministry of the Environment (2001).

environmentalists and Water Agency officials, water charges serve a
resource conservation purpose, and should be only devoted to the
improvement of quantity and quality of water resources for the sake of
all users. For government officials, however, water charges are similar
to a fiscal instrument, of the same nature as say a tax on company
profits, and should be freely usable for other purposes, for example
youth employment policies. The Conseil d'Etat (French equivalent of
the US Supreme Court) stipulated in 1982 that water charges were
actually comparable to a genuine tax and could therefore be included in
the general budget of the state.

The original purpose of the project was to have a better control of
public funds raised by Water Agencies, but internal political
considerations also played a role in the motivation for this project.
Supporters of the 'double-dividend' policies argued that since water
charges could now be considered a genuine tax, their revenue could
be used to fund the government project for youth employment.
This project was abandoned before the 2002 general elections, but
given the significant level of tax revenues collected by Water Agencies
it is likely that it could again become an interesting option in the
future.

Revenue distribution
According to the Water Act of December 1964: 'Water Agencies should
grant loans and subsidies to agents engaged in projects of common
interest to the river basin, the outcome of these projects leading to a
reduction in the future financial expenditures that otherwise should be
incurred by agencies.' So, Water Agencies should finance projects

whose positive externality in terms of quantitative and qualitative water resource management would benefit all river basin users and stakeholders. As remarked by Salanié and Thomas (1995), without the intervention of Water Agencies, these projects would be executed in an inefficient way or would not be executed at all, since individual agents would not take into account the externality issue in their decision-making process. By this procedure, financial transfers by Water Agencies should rationalize the intervention of all agents in order to reduce the collective costs.

In practice one can observe that the eligibility criteria have not been applied to meet explicit environmental targets as a way to achieve the 'minimum social cost'. As a recent report by the Commissariat Général du Plan (2001) remarks: 'Nowadays, an agent is eligible to receive financial transfers by the sole criterion that its project is in the scope of the working plan of the Water Agency and there are financial resources available to the corresponding intervention guideline.'

Thus qualitative reasoning has been replaced by budgetary reasoning. This current eligibility criterion has the advantage of simplifying the agencies' actions. On the other hand, important issues are ignored, sometimes at the expense of the rational resource management. Three observations can illustrate this argument:

1. Small agents are not targeted by Water Agencies. WAs prefer to direct subsidies toward large users or polluters, even when this is not the most efficient policy. For example, Agencies prefer to finance collective sewage systems rather than individual ones, even when it is proved that in some regions the latter would be more adequate in terms of environmental performance. When dealing with individual agents, Water Agencies prefer to use unions or local communities as intermediaries, which are in charge of transferring the funds. This may prevent an efficient allocation of Agencies' funds.

2. Absence of project managers prevents Water Agency's intervention. Since the WA can only grant subsidies to projects once a responsible agent for the project can be identified, problems such as non-point source pollution cannot be handled properly.

3. Private concessionaires of water distribution services can be found among Water Agencies' direct beneficiaries. These agents are supposed to be in charge of their equipment and its maintenance. By receiving subsidies from the Agencies, they have a double benefit: firstly, by reducing their investment cost; secondly, the

projects are often conducted by concessionaires' subsidiaries, creating new market opportunities for the company.

Table 3.8 presents the distribution of subsidies during the sixth (1992–96) and seventh (1997–2001) Water Agencies' working plans. The first point that becomes evident is the priority given to pollution control abatement. In both working plans, pollution control represents 83 per cent of the total subsidies, while water resource management accounts for the remaining 17 per cent. One can also observe that Water Agencies favour construction and civil engineering projects. As Table 3.8 shows, financial support to operation and maintenance costs (water treatment premium plus operational cost subsidy) are small if compared to the construction of treatment plants and sewage networks, even if the former have recently gained in importance, as shown by the rates of change figures. Finally, the emphasis put on sanitation works reflects the importance of European directives in the French water management system. This is a good pattern in environmental terms but it is hard to claim that the 'polluter pays' principle is really being put into practice.

3.5.7 Compensatory Measures and Distortionary Fiscal Instruments

Some macroeconomic, regulatory or local policies may, and sometimes, did affect the performance of water charge implementation. However, most of these distortions originated from the 'command-and-control' side of industries and from the agricultural policy for farmers. For industries, as production serves in most cases as a proxy for computing emission and water use charges, an increase in emission charges following a rise in production output can be partly compensated for by fiscal policy oriented toward employment or regional industrial development. Macroeconomic or sectoral policies having as objectives the preservation of regional employment have helped industries to develop or sustain themselves in less-favoured regions, with direct consequences on the environment.

For instance the government has since the 1980s designed policies oriented toward traditional, labour-intensive industries, for subsidizing low-skilled employment. However, industries concerned with this category of workers were also in most cases associated with high emission rates, and found it difficult to continue operations given the effluent emission charges collected by the Water Agency. In this case

Table 3.8 Subsidies by type of operation (in million French francs)

	VI Working Plan	% of total subsidies	VII Working Plan	% of total subsidies	% change
POLLUTION					
Treatment plants in communities	10864	25	12915	23	19
Sewage network	11392	27	13424	24	18
Industrial pollution control	5949	14	6048	11	2
Waste disposal	1159	3	1178	2	2
Technical assistance	370	1	631	1	71
Water treatment premium	4730	11	7980	14	69
Operational costs subsidy	614	1	2189	4	257
Agricultural pollution control	550	1	2682	5	388
Others	42	0	169	0	302
Total	35652	83	47216	83	32
RESOURCE AVAILABILITY					
Waterworks	815	2	1114	2	37
Irrigation	161	0	25	0	-84
Groundwater	726	2	643	1	-11
River basin recovery	711	2	1548	3	118
Drinkable water	4469	10	5520	10	24
Resource management	393	1	892	2	127
Total	7275	17	9742	17	34
Grand total	42927	100	56958	100	33

Source: Ministry of the Environment (2001).

the distortion has exactly the effect of a 'double negative dividend': public funds are devoted to employment policies, which in turn favour more pollution.

In regard to the WA policy for subsidizing pollution abatement capital cost, it has been acknowledged that it alleviates the financial burden of water charges for industries and local communities. Many analysts describe subsidies for abatement fixed cost as an indirect way to compensate polluters. For polluters not engaged in pollution abatement activities, such subsidies clearly do not exist, but this should concern only the fraction of industrialists with limited levels of effluent emissions anyway. Abatement subsidies partly compensate for water charges paid for water use and/or effluent emissions (respectively, new water-saving production processes and abatement facilities). But viewing the water quality management system as a whole, these subsidies also compensate for the cost of compliance with effluent emission standards. As a consequence the WA subsidy policy can be seen as a compensatory measure creating a distortion in the impact of water charges of the very same WAs.

3.5.8   Identification of Major Drawbacks in the French Water Agency System

The first obvious drawback of the system is that a major source of effluent emissions as well as the most important contributor to total water net consumption is virtually exempted from the charge system. Farmers (with the exception of some animal breeding activities) are currently not subject to emission charges that would correspond to their nitrogen and pesticide pollution. Hence we will concentrate our analysis on residential and industrial users who have been participating in the charge system since its beginning.

Residential users
The main criticism of the French effluent emission charge is that it does not give an accurate price signal to the consumer on the cost of abatement following his or her water consumption. First, the charge is based on consumption instead of pollution levels. For example one household may use a given volume of water for gardening while another one uses the same amount for washing clothes. Both households will pay the same charge for effluent emission even if the latter's emission is much more damaging to the environment.

Furthermore the effluent emissions charge is based not on real consumption but on expected total water consumption. It is said that this design leads to cross-subsidization between polluted and polluters. As the total expected consumption is divided equally between all consumers, the ones who pollute less are paying for the biggest polluters. Ideally the charge should be a function of total effluent emissions by the user. As we know, it is not possible to measure the pollution generated by each consumer because wastewater is collected and treated in a single abatement plant for the whole local community.

This is why it has been often argued that charging the local community instead of the user would be more relevant. Indeed, even if the residential user was informed about the charges paid for environmental preservation, he would not be able to modify the pollution level generated by his activities. In other words the type and amount of pollution depend on the sewage facilities of the community and not directly on consumer actions. Local communities that chose an efficient sewage system would be rewarded by a lower charge. This charge would then be shared equally between residential consumers. Of course rules for sharing the total cost of emission charges would have to be planned if two or more communities use the treatment plant jointly.

The effluent emissions charge, based on expected total consumption, is thus more a flat and broad measure of pollution. Figures on net consumption as well as statistics on effluent emissions are not well known and often differ significantly between sources. It is also the case that the Water Agency does not have an exhaustive record of treatment plants belonging to its zone. This is the case when a private firm in charge of water management in a community asks another private company to build the treatment plant. Agencies are not perfectly informed on the number of connections to the sewage networks either. This lack of information is exacerbated because there is no central institution that collects and gathers data on sanitation in France.

Two other coefficients used for the computation of the final charge rate have also been heavily criticized on equity grounds. A coefficient depending on population size was introduced in order to account for the lower coverage levels of household sanitary and other domestic facilities (washing machines, for example) in small rural communities. Today this differentiated treatment is no longer justified, given the widespread presence of domestic facilities in both rural and urban communities. As a result, the coefficient represents a benefit to rural communities. The water charge formula also includes a coefficient for effluent collection, created in 1992 to finance building of sewage networks. This coefficient, which amounts to multiplying previous water charge values by a factor of 2 or 3, just applies for residential users. This is not seen as an equitable mechanism since industrial effluents can also be discharged in the sewage network.

Even if it is difficult to get accurate information on the cost of abatement, and also to determine whether actual charge rates are close to the optimal ones or not, experts agree on the too-low level of the rates (charge per cubic metre). Effluent charges are on average about 30 per cent of sewage payments, but effects on user's behaviour are constrained by the above-mentioned cross-subsidy effects that mistarget the effective polluters. Apart from that there are also cost-recovery and political constraints, as follows.

Firstly, Water Agencies cannot freely determine the charge level. Since the 1964 Water Act, the level of the charge is constrained by the total amount granted to the Water Agency for the corresponding five-year working programme. So, cost-recovering ceilings limit price incentives.

Secondly, since the 1992 Water Act, the budget for water supply and sanitation has to be set apart from the general budget of the local communities. As a consequence, the budget for water has to be balanced: users' payments have to cover all costs implied by the distribution and treatment of water. However, as the Water Agency will cover part of the investment cost (through subsidization), not all the cost will have to be borne by the consumers.

Thirdly, the price of water has increased dramatically since 1992, and residential consumers are now more sensitive to their water bill. Residential users in some towns have refused to pay their bill. It is more difficult, in political terms, for politicians and water utility operators to let the price of water increase even further without sound justification.

3.5.9 Industrial Users

An important review by environmental and fiscal experts was conducted in November 1993, on the whole Water Agency system. Their report identified the following problems.

There is a lack of transparency and equity in the effluent emission charge system. Geographical disparities between Agencies for the effluent emission unit charges range between a factor of 3 and 7 for BOD, and between a factor of 1 and 9 for phosphate. Also, industrialists may see the emission rates per unit of output modified by Water Agencies depending on the result of self-reporting, or following random external auditing that sometimes shows some lack of transparency.

Without questioning the autonomy of Water Agencies, government officials claimed recently that environmental objectives were to be decided at the national level, following the new Water Act of 1992. Consequently priorities within the framework of the national quality control policies should be decided upon at the national level, and then implemented in practice at the local level by Water Agencies.

Since industrial payers can recover part of their effluent emission charges through subsidized abatement facilities, it is necessary to define a new approach for the integrated water management at the river basin level. But this also raises the issue of the abundant regulatory components of water management in France, and revision of this approach would also require simplifying command-and-control oriented policies.

Another major concern is the lack of institutional basis for data collecting, processing and analysis at the national level, by which trends in pollution and water charges could be analysed in a consistent and continuous way. The National River Basin Network (RNB) collects information on water quality parameters for about 1000 measurement stations, but accidental pollution and short-term variations in effluent emissions are difficult to detect with such a system in its current state. Furthermore no attempt has been made to merge this measurement network with the Water Agency databases for water charges in order to capture the expected impact of water charges on overall effluent concentration. For this reason there is no database containing both discharge and actual effluent concentrations data, at a national or even at a river basin level.

At the national level, the measurement of effluent concentrations yields very different results according to the source of study, with major discrepancies for suspended solids and BOD as measured either by Water Agencies or by the governmental offices for environmental protection and human health. A major explanation for this is the fact that not all pollutants are charged by every Water Agency. Indeed because of the decentralized nature of the system, Water Agencies have a tendency to charge effluent emissions that are most problematic in their own river basin (for example, salt in eastern parts of the country) and overlook others.

The major source of heterogeneity among effluent concentration measures is due to the very nature of the charge collection process by Water Agencies. As noted above, the system of effluent emission charges made proportionally to output has been one of the reasons for the early success of water charges in France, due to its relatively low cost of implementation. However, with the incorporation of new pollutants and the change in the industrial structure over the past decades, this system is becoming insufficient and increasingly less consistent with the necessity to keep track of environmental impacts (that is, actual impact versus theoretical ones).

There is the same problem with the measurement of water use. Comparing the water consumption levels charged by water utility operators with the levels charged by Water Agencies, discrepancies of up to 70 per cent have been computed recently.

3.6 RECOMMENDATIONS

3.6.1 The Role of the Water Agency

Clearly, designing the role of the Water Agency is a central issue. Should it play the role of a bank or a mutual company, or should it be a public organism managing the charge and rebate system by itself? Should the Agency be allowed the possibility to design the level of the charge on its own? Today in France, WAs have no clear guidelines for the computation of the charge (which pollutant to target, which coefficient to apply and so on). The level of the charge is constrained by the amount allocated to each Agency for a five-year programme. Moreover no continuous evaluation process is in place to analyse the effects of the charge system on uses and on the environment. As a consequence Water Agencies find it difficult to decide between competing projects and to compare their potential expected outcomes. Not surprisingly, almost all projects that are submitted are funded (for example, through corresponding subsidy programmes), with rebating representing an average 30 per cent of the total cost of the project. Although that can be the easiest way for the Agency to keep maximum political support for the Agency system, it also increases the possibility of collusion between the Agency and local users and polluters.

More frequent measures of pollution and a continuous evaluation of the programmes (returns in terms of avoided pollution, impact on environmental quality, socio-economic consequences) have to be encouraged. As mentioned before, governmental authorities for environmental protection are also in charge of emission monitoring for the purpose of law enforcement. The question therefore remains as to which authority (governmental authority or the Water Agency) would be most suitable to engage in more frequent emission measures. Since Water Agencies provide financial support for operations concerning water use and wastewater treatment activities by industrialists and local communities, Agencies seem to be in a better position to conduct such improved measurement campaigns. This, however, raises the issue of the lack of information-sharing between the Water Agencies and the environmental protection agency, as the latter could benefit from more frequent emission measures by Water Agencies.

3.6.2   User Information

As noted above, there is a need for more transparency and equity considerations in the effluent emission and water use charges. The residential user is not perfectly informed on the charges collected by the Agency because charges are not always easily observed in the water bill. It is probably the case that some residential users do not even know the existence of a Water Agency, as the local supplier (water utility operator) plays the role of an intermediary between users and the Water Agency. This kind of user information through bills is essential for the success of policies devoted to promoting more efficient water use. In addition to that, information to residential users could also be improved if users' participation in Basin Committees was strengthened, since they are proportionally the less-represented user group, and due to their diversity, also the less-organized one.

For industrialists and farmers, since they gain directly from rebates, exemptions and all kind of subsidies, information on water charges is less of a problem. Nevertheless more systematic information should be made available on the impact of their activity on the environment and how investments funded by the Agencies are performing.

3.6.3   Design of the Charge

Water Agencies need more detailed guidelines for designing the effluent emission charge. WAs should first agree on the potential damage of pollutants, based on scientific and technical auditing reports, and then decide on a common list of targeted pollutants. When this is done, WAs should agree on the type of coefficients to apply for computing charges, while each Water Agency would of course be allowed to decide on the level of the unit charge. The formula should be the same and the coefficients modulated depending on the sensitivity and local characteristics of the zone. This would guarantee equity between basins, but also equity between users by making all polluters liable, and so suppressing coefficients of use types. Such an approach would not lead all Agencies to target the same pollutants, but rather the most damaging ones in all circumstances and environments.

Moreover, as observed by Gastaldo (1994), the lack of a genuine application of the 'polluter pays' principle implies that environmental objectives cannot be reached when unit emission charges are not offering price incentives for use pattern changes. Since the recent trend has been toward significant increases of these rates, scientific evaluations are needed to determine how the magnitude of these increases is affecting environmental quality.

3.6.4 The Need for a National Water Committee

As mentioned before, the lack of a centralized institution for data collecting, processing and analysis at the national level makes difficult the assessment of the environmental and economic outcomes of the charge system. With the growing predominance of European directives in the design of national environmental policies, environmental objectives will be increasingly decided at the national level. Hence priorities for quality control and water management should be designed at the national level and then implemented at the local level by Water Agencies.

Therefore there is a need for a 'national water committee'. This institution could also coordinate efforts among all regulatory forces (water police, regional directions for industry and environment) and Water Agencies to simplify and consolidate their actions, avoiding redundancy and conflicting targets and instruments.

3.6.5 Towards a More Sustainable Policy Implementation Path

Many problems in the way the French Water Agencies operate have already been identified. However, it should be remembered that such a system was implemented more than 30 years ago, when very few past experiences on charge systems were available, and with sparse scientific evidence on water environmental impacts. So the most important point to stress is that it is crucial to have a realistic view of what the final objective of the projected water charge system should be. More precisely, what are the pollutants to consider for abatement and what should be the final water quality objective. Next, potential institutional and legal reforms must be designed to remove implementation constraints. In this respect one can support the view that the French way of implementing the Water Agency system is interesting to consider from an historical perspective, as it consists in several stages, each one being associated with different knowledge and public perception. Based on the French experience we can suggest two development phases for a water charge system.

In the first stage, a system of effluent emission and water use charge is implemented using fixed proportion coefficients between production, use and emission levels. Only a subset of major pollutants is considered for taxation and the most important users (in terms of size of activity and ability to provide pollution and use data) are charged. Although this is inefficient from an environmental viewpoint, it would allow water users to be familiar with the charge system and to learn more about the way their behaviour can be modified to accommodate higher charge rates. Hence the main objective in this first stage is primarily to collect information on pollution levels by regions and to have a preliminary notion of expected fiscal revenue from charges.

In the second stage, the set of pollutants is widened to cover other important substances, including agricultural non-point sources. Unit charges are gradually increased depending on final environmental objectives and public reaction resulting from the first implementation stage.

This way of implementing a consistent economic instrument-oriented policy for water quality has the advantage of imposing only a moderate financial burden (in the form of water charges) on users for a limited period of time, and to avoid severe consequences on employment and economic activity if high charge rates were decided upon at the first implementation stage. Of course the risk is to let the first stage pertain for a long period of time, overlooking the final environmental objective. The length of the first stage would depend not only on users' reaction and first environmental outcomes, but also on political considerations. Ultimately the most important political issue concerns the trade-offs between the price incentive nature of the charge system and the need for subsidizing users and polluters.

NOTES

1. This chapter was part of a series of papers commissioned by the Inter-American Development Bank for the Environmental Policy Dialogue and the opinions expressed in this chapter are solely those of the author and do not necessarily reflect the position of the IADB.
2. See Figure 3.1, p. 56.
3. Sample 1: 116 water utilities from eastern France between 1988 and 1995. Sample 2: 108 water utilities from the south-west of France between 1990 and 1994 (see Nauges and Reynaud, 2001).

REFERENCES

Amigues, J.-P., M. Moreaux, F. Salanié, A. Thomas and Q.H. Vuong (1994), 'La régulation de la pollution industrielle par les Agences de l'Eau' ('The regulation of industrial pollution by Water Agencies'), Paris: Commissariat Général du Plan.

CGGREF (Conseil Général du Génie Rural, des Eaux et des Forêts) (1994), 'Evaluation of the Sixth Program of the Water Agencies', Paris: Ministry of the Environment.

Commissariat Général du Plan (2001), 'La politique de preservation de la resource en eau destinée à la consommation humaine' ('The water conservation policy for human consumption'), Paris: La Documentation Française.

DGCCRF (Direction Général de la Concurrence, de la Consommation et de la Regression des Fraudes) (2001), *Enquête sur le prix de l'eau 1995/2000* (Water Price Survey, 1995–2000), Paris: Ministry of the Economy.

European Community Council (1991), Directive on the treatment of urban wastewater, 91/271/CEE, Brussels.

Gastaldo, S. (1994), 'Les instruments économiques à l'appui des politiques de l'environnement' ('Economic instruments for environmental policy'), Paris: Direction des Etudes et Synthèses Economiques, INSEE.

Guide de l'Eau (2001), Paris: Johanès Editions.

Lavergne, P. and A. Thomas (2004), 'Semiparametric estimation and testing in models of adverse selection, with an application to environmental regulation', forthcoming in *Empirical Economics*.

Martin, Y. (1996), 'La gestion durable des eaux souterraines en France', Paris: Conseil Général des Mines.

Ministry of the Environment (2001), 'Report on the Environment', Paris: La Documentation Française.

Nauges, C. and A. Reynaud (2001), 'Estimation de la demande domestique d'eau potable en France', *Revue Economique*, **52** (1), 167–85.

Nauges, C. and A. Thomas (2000), 'Privately operated water utilities, municipal price negotiation, and estimation of residential water demand: the case of France', *Land Economics*, **76** (1), 68–85.

Neira, M. (1995), 'Etude du système des redevances des Agences de l'Eau en France' ('Analysis of the French Water Agency Tax System'), report for the Rhin-Meuse River Basin Committee, Metz.

Salanié, F. and A. Thomas (1995), 'Impact des redevances sur les équipements d'épuration et analyse des politiques d'aide à l'investissement' ('Impact of emission charges on abatement and an analysis of abatement subsidy policies'), report for the Water Agency Committee, Toulouse.

4. Country case: Mexico[1]

Lilian Saade Hazin and Antonio Saade Hazin

4.1 INTRODUCTION

Since the establishment of a central agency in charge of the use of federal water resources in 1989, water management in Mexico and its regulation have been reformed to improve its efficiency and effectiveness. The process started with a system dependent on governmental budget allocations and for the most part focused on irrigation-related investments, and moved towards a more market-oriented scheme addressing the multiple needs of a growing urban population.

The water supply system as a whole remains highly subsidized. Household water tariffs do not cover direct costs. Agriculture is by far the major user – with almost 80 per cent of the total water supplied to the system – and farmers get water at no cost, except for the electricity it consumes to pump it.

As part of the strategy to modernize the water sector, the Mexican authorities have endorsed the use of economic instruments as one of the main approaches. Following this line of thought, the Environmental Program for 1994–2000 declared that: 'economic instruments present advantages that make them attractive and necessary for public policies in environmental stewardship'. The water use charge from federal water bodies put into operation since 1986, and the wastewater charge implemented in October 1991 are the main examples of these instruments.

After this introduction, the following section describes water problems in Mexico. The third deals with the institutional and legal frameworks for water management in Mexico. The fourth section describes water use charges in Mexico and the measures for wastewater pollution control, the fifth section includes the difficulties in the implementation of the policy instruments, revenue issues and the financing schemes in place. Finally, conclusions are drawn.

4.2   POLICY PHASE

This section describes the policy background of water charges in Mexico by identifying the main water problems and the evolution of the institutional and legal frameworks.

4.2.1   Water Problems in Mexico

The main water problem in Mexico is not the average availability of the resource. Water availability per capita is 5000 cubic metres a year. This is more than the amount of water available for each person in France or Spain; and many times more than in Egypt. However, around 89 per cent of the population has access to drinking water in Mexico, a figure clearly smaller than in those countries (CNA, 2002b).

The real water management problem in Mexico is the distribution of water resources. Economic activity and population density by regions do not correspond to the location of the country's rivers and aquifers, which makes distribution difficult and costly. Thus 76 per cent of the population lives in the northern and upland region of the country, which has 20 per cent of the water resources. Most of these populated areas are arid. A number of regions present serious availability problems. To aggravate this situation, one-quarter of the population lives in areas that are 2000 metres above sea level.

In this context, overexploitation of aquifers in relative scarcity zones is a growing concern. In 1975, there were 32 overexploited aquifers. In 1981, the number was 36, in 1985, it increased to 80 and in the year 2000, it reached 96 (Semarnap-CNA, 2001).

The situation is rapidly deteriorating. The accelerated population growth in the past 50 years has reduced water availability per capita in Mexico from approximately 18 500 m^3 per capita in 1950 to less than 5000 m^3 today (Tecsasím-Lyonnaise des Eaux, 2000). At the current population growth rate, the total water availability or water quality will critically decrease in the densely populated and more industrialized regions. With respect to sewerage, 77 per cent of the Mexican population has access to sewerage services.

Water pollution has been an increasing economic problem and a main health hazard in Mexico. Despite the importance of the need to address this relative scarcity, water treatment remains low. Only 5 per cent of the 535 surface water bodies in the country are classified as excellent and can be drunk without treatment. An additional 22 per cent is considered 'acceptable', but needing some degree of treatment to be drinkable. The remaining 73 per cent represent more acute pollution problems to different degrees (CNA, 2002a).

4.2.2 Institutional Framework

The Mexican legislation considers water as a national resource. All major water bodies in Mexico are a matter of federal responsibility. There are minor bodies in charge of state authorities, when a superficial body is entirely within the boundaries of a single state territory. The same applies to superficial streams and water bodies that are confined to a single property and are considered as an integral part of the land.

A central agency in charge of the use of federal water resources, the National Water Commission (CNA) was created in 1989. The CNA is the sole authority for federal water management and is in charge of the promotion and execution of federal infrastructure and the necessary services for the preservation of water quality.

The CNA is attached to the Ministry of Environment and Natural Resources. Prior to this it was attached to the Ministry of Agriculture and Water Resources, which historically had retained the main authority over matters of water control and federal investment. The new structure and its institutional relationship makes evident the environmental concerns of the Mexican authorities with respect to water issues.

Regardless of the change in institutional structures, the main goals of the CNA are not environmental protection or regulation, but infrastructure construction and operation. It can be said that it is more a 'rowing' than 'steering' institution. The most important part of the budget is oriented towards financing infrastructure, particularly in agriculture. In the year 1991, more than 37 per cent of its budget was destined to pay for irrigation infrastructure and 33 for water and sewerage works. The remaining 30 per cent was used for general management and regulation-related activities. Only a small fraction of this percentage was assigned to monitoring.

Under the new arrangement, the federal water management system encompasses 13 administrative regions defined by the CNA, following hydrographic criteria. Each region comprises one or more basins, thus basins and not states are the basic division of the Mexican water management system.

Close to the French experience, the system counts on 26 River Basin Councils. Following the French principles, the objective of the Councils is to promote social participation, taking into account the opinion of every stakeholder involved. They contribute chiefly to the planning of water policies. Despite its importance, the current law does not detail the characteristics of the Councils or their functions; it only mentions their existence as a participatory body for the CNA, the local authorities and water users. In practice, the Councils have not been fully implemented and their capability has not been completely developed. Up to 1997 only one such Council was established. By 2000 a total of 25 of them had been created. In addition, six Basin Commissions and four Basin Committees have been established. These only differ in terms of the size of basin they cover. They are not agencies in their own right, but catchment-based coordinating forums bringing together water users and government organizations at different levels.

At a decentralized level, municipalities are in charge of water, sewerage services and wastewater collection and disposal. However, due to the weakness or lack of capacity of most municipal operators and some legal restrictions that apply to the approval of water tariffs, state authorities have an important role.

4.2.3 Legal Framework

In chronological order, the first important legal water text was the 1910 Water Law. This law declared national waters to be in the public domain and for common use. The 1926 Federal Law of Irrigation Waters regulated the government efforts to promote and invest in irrigation infrastructure. The particularity of the Law of Irrigation Waters was the creation of a construction permit. It was replaced by the Law of National Waters Property of 1936, which detailed the awarding of federal concessions and regulated user associations (Brañes, 2000). Other water-related laws followed to regulate sanitation infrastructure and underground water uses.

In 1972 a revised Federal Law of Waters was approved. As with the previous laws, it emphasized the status of national property over water bodies mandated by the Constitution and created a centralized system of permits and concessions for water use. This law was the basis of the first National Water Plan in 1975, which gave a unified vision to the issue.

The allocation of resources as mandated by the 1972 law was far from efficient. The concessions granted for water use did not only reflect economic considerations, but also political, social or simply bureaucratic ones. More importantly, the supervising structure lacked the resources needed and was unable to monitor effectively the proper use of water, which eventually led to abuses. In 1988 the excessive misuse of water forced the authorities to temporarily forbid the exploitation of 55 per cent of all water catchments in the country until they returned to acceptable levels (Roemer, 1996).

The following and most recent regulation was the 1992 National Water Law. It is the backbone of the federal water system, but not the only law linked with water regulation. These include a Law of General Ecological Equilibrium and Environmental Protection, a General Health Law, a Federal Law of Charges, a Law of Water Works and a Fisheries Law.

In particular there is a conflict of competence between the environmental and water authorities, conflict that leaves a grey area between the Ministry of Environment and Natural Resources and the CNA. A projected law entitled Ley de Cuencas y Aguas Nacionales (Basin and National Water Law) and at least two other projects to reform the current law are being analysed by the Congress. While it is not certain that any of these will be approved, it is worth mentioning that all of them propose to detail the functions of the Councils in the law, seeking to empower them. Certainly these proposals offer only limited improvements in terms of institutional autonomy and decentralization, but it is an undeniable goal of all of them to strengthen the institutional capacity of the participatory organisms.

4.3   DESIGN PHASE

This section analyses the evolution of the water system in Mexico since the amendment of the National Water Law in December 1986 when the use charges first included an increasing block rate structure.

4.3.1   Water Use Charges

The Mexican legislation contemplates water use charges as well as wastewater effluent charges. In principle the National Water Commission is allowed to charge a fee to water users for the right to use the federal water bodies. The amount collected in water use charges accounts for more than 53 per cent of the budget of the CNA (CNA, 2002a), although collection was expected to drop at the end of 2002.

Before 1986 the pricing system employed one fixed price per cubic metre throughout the country. In 1986 the Federal Law of Charges was modified and a two-part charge came into place: one part was a fixed price per cubic metre of water used, but varying depending on the water supply zone. The other part was an increasing block rate structure. In addition, four major water uses were established – irrigation, hydro-electric generation, urban (potable) and industrial – and each use was assigned a pricing weight. Table 4.1 shows the water use charge rates in 1993.

Table 4.1   Water use charges ($US/m³)

1993	Industry	Urban (potable)
Zone 1 Scarce	0.419	0.193
Zone 2 Equilibrium	0.290	0.009
Zone 3 Sufficient	0.103	0.005
Zone 4 Abundant	0.077	0.002

Source: Guerrero and Howe (2000). The exchange rate used is from INEGI (2002).

In 1997 the CNA changed from four to nine the number of availability zones. The change in zones – previously defined from a hydrological point of view – was based on specific administrative concerns. Under the new scheme the former zone one (the scarcity zone) was divided into six zones. The division reflected the existence of different policies and practical criteria for the application of subsidies, permissions and exceptions to industry in some regions. The rest of the zones simply changed their number. Zone seven is equivalent to the previous zone two (equilibrium), the zone eight to the former zone three (sufficient), and zone nine is equivalent to the former zone four (abundant) (Guerrero, 2002).

In its actual form, the federal water use charge depends on two factors: the availability zone and the use that is going to be given to the resource. Table 4.2 reproduces the rates published with the 2002 Federal Law of Charges in US dollars at a rate of ten pesos per dollar. Each year the law updates the charge level, and sometimes a region can change availability zone according to its most recent hydrological balance. Some of them have moved from zone 9 to 8 for example. As can be seen in the 2002 charge arrangement (Table 4.2), only industrial use exhibits a difference among zones one to six, as charges for the other uses are identical between the first six zones.

Table 4.2 2002 water charges ($US/1000m³)

2002	Industry (general case)	Urban (potable)	Hydroelectricity	Aquaculture	Recreation
1 Scarce	860	19.95	0.18	0.12	0.49
2	690	✓	✓	✓	✓
3	570	✓	✓	✓	✓
4	470	✓	✓	✓	✓
5	370	✓	✓	✓	✓
6	340	✓	✓	✓	✓
7 Equilibrium	250	9.29	✓	0.06	0.24
8 Sufficient	90	4.64	✓	0.03	0.11
9 Abundant	70	2.32	✓	0.01	0.05

Source: Federal Law of Charges (2002).

As also shown in Table 4.2, public and private entities that serve for drinking water must pay lower rates than industrial users. Furthermore the operators working for communities with less than 2500 inhabitants are exempted from the charge. Other exemptions include:

- water for agricultural use (almost 80 per cent of the total water used)
- wastewater
- water with high concentrations of salt in interior water bodies
- users holding a certificate issued by the CNA indicating that the water was returned to its source without change

To encourage wastewater treatment, a special discount is due when the user treats the water before returning it to the source and the treatment complies with the quality standards established.

The water use charge structure reflects two main objectives. The first is to achieve better economic efficiency in the use of water, charging more when the resource is scarce and less when it is relatively abundant. The second objective is to promote certain uses that indicate the policy priorities of the federal government, such as the satisfaction of the basic needs of the urban population. A net effect that can be seen in Figure 4.1 is that industrial use makes the greatest revenue for the CNA, even when municipal water systems consume more.

4.3.2 Wastewater Pollution Control

Wastewater quality control in the 1970s called for a basic treatment for any discharge, regardless of the type of water-receiving body. Due to investment problems and, particularly, to the lack of monitoring systems, this scheme failed.

During the 1980s and the mid-1990s public policy emphasis was placed on controlling wastewater from industries with a high organic load or toxicity. Recognizing the failure of the environmental and water authorities to monitor all polluters, it was considered satisfactory to control only the main discharges.

In 1991, a wastewater charge was put into operation to make the polluters internalize the costs of their actions, providing them with an incentive to comply with the pollution control regulations, resting on the 'polluter pays' principle. Since its inception, the priority of the wastewater charge has been to induce polluters to comply with the

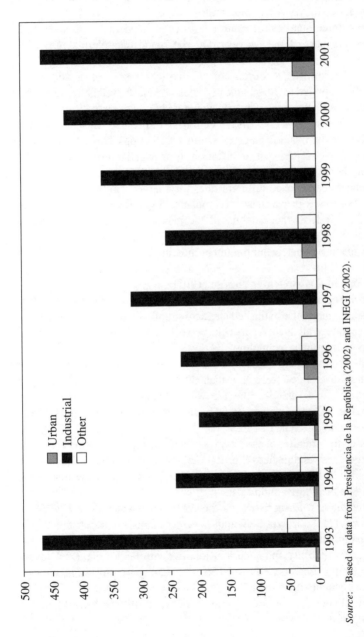

Source: Based on data from Presidencia de la República (2002) and INEGI (2002).

Figure 4.1   CNA revenue for water use charges (millions of $US)

regulations, rather than to increase the revenue. Polluters only pay for units above the discharge standard.

To enforce that kind of control, a series of standards were elaborated. These are part of the Mexican Official Standards (Normas Oficiales Mexicanas, or NOM). The parameters used for the standards were determined according to the typical result of a secondary treatment applied to a discharge of a specific industry sector, regardless of the characteristic of the water-receiving body (Giner de los Ríos, 1997).

As discussed in the next section, the original mechanism has undergone two major changes worth mentioning. The 1995 revision changed the criteria of application from availability zones to the 'assimilative capacity of the water bodies'. Finally, the 1997 version of the charge established different rates for a larger number of pollutants. The 1997 charge was also accompanied by a review of the official quality standards that established the maximum permissible levels of pollutants for wastewater discharges, and new rules to promote investment in wastewater treatment infrastructure.

The original wastewater charge (1991)
The charge was a non-compliance charge applied when the pollutant units exceeded the established standard. Implicitly the idea of the charge was to induce all agents, particularly those in highly polluting activities, to adopt treatment technology.

The original charge applied a flat rate – related only to volume levels – when discharges were less than 3000 m³ a month. For discharges exceeding that quantity, the fee was calculated on the base of the chemical oxygen demand (COD) and total suspended solids (TSS) released and the type of zone.

Since the creation of the first wastewater charge, the Federal Law of Charges has established a procedure for self-reporting. The user is responsible for measuring the volume and contents of the discharge, calculate the charge and pay it.

The main characteristics of the 1991 wastewater charge were as follows. (1) Four types of zones were established according to the availability of the resource. Zone one corresponded to regions with the highest availability and zone four to areas with the lowest. (2) The rate varied depending on COD and TSS. The total charge would be the sum of the elements in Table 4.3.

The Federal Law of Charges conceded the following charge exemptions:

Table 4.3   Level of charges

Zone	For every 1000 m³ of wastewater US$	For a ton of COD US$	For a ton of TSS US$
1	78.70	51.16	90.50
2	19.64	12.70	22.60
3	7.77	5.08	9.00
4	3.91	2.50	2.50

Source: Based on the 1991 Law of Charges and INEGI (2002).

- users who complied with the effluent standards
- users holding a certificate issued by the CNA indicating that the water was used and returned to its source without change
- users discharging into non-federal bodies or sewer systems

The charge is applied by the CNA and, in its original version, the resulting revenue went to the Treasury without returning to either the municipalities or polluters.

The 1996 Federal Law of Charges
In 1996 the Federal Law of Charges was amended. A new procedure was set to calculate the charge, no longer based on availability zones but calculated according to the assimilative capacity of the different types of receiving water bodies. Three broad categories of water bodies were created:

1. water bodies requiring a lower treatment level
2. water bodies that require secondary treatment
3. water bodies that require a more sophisticated treatment level

The charge rate also varied according to COD and TSS concentration levels. As a change from the previous version, the charge would be calculated applying a fee only for the pollutant that was more concentrated according to the charges in Table 4.4.

The Federal Law of Charges conceded the following charge exemptions to:

Table 4.4 Pollutant concentration

Type of water body	Pollutant concentration			
	Less than 30 mg/l US$	30–75 mg/l US$	75–150 mg/l US$	More than 150 mg/l US$
A	0.04	0.04	$(0.01 \times$ concentration$) - 0.69$	0.77
B	0.09	$(0.07 \times$ concentration$) - 0.47$	0.77	0.77
C	$(0.07 \times$ concentration$) - 0.47$	0.37	1.60	1.60

Source: Based on the 1996 Law of Charges and INEGI (2002).

- users who complied with the effluent standards
- users holding a certificate issued by the CNA indicating that the water was used and returned to its source without change
- users discharging into non-federal bodies or sewer systems
- public water suppliers to municipalities of less than 2500 inhabitants
- discharges from agricultural irrigation
- users with an approved programme to reduce their polluting emissions

Polluters with a monthly discharge volume of less than 3000 m³ had the option of a flat rate.

From 1997 onwards
Additional pollution indicators are now considered in a stricter and more complex formula. Charges are again calculated taking into account a particular tariff for every pollutant in the water, rather than a payment for the one that is more concentrated, as in 1991.

As an important innovation, the 2002 Federal Law of Charges determined that the amount obtained from the charges from public or private firms will be destined for activities for improving efficiency and water-related infrastructure. Before the introduction of this change, the money collected was diverted to the Treasury. Therefore, CNA had little incentive to collect revenues. As long as the resources were not necessarily kept in the sector, payers did not see the benefit.

To summarize the changes, Figure 4.2 presents a diagram of the process as the 2002 Law establishes it.

Industrial discharges
Although water abstractions for industrial use represent one-tenth of total abstractions, industrial discharges are a particularly important issue given the quantity and diversity of pollutants discharged by some industries. Industrial discharges represent around 170 m³/s of waste-water. This is more than 6 million tons of BOD, which surpasses by 140 per cent the pollution generated by urban water and sanitation services (Semarnap-CNA, 2000).

As can be seen in Table 4.5, among the main industries generating wastewater discharges are sugar and chemicals. There are 1485 industrial wastewater plants of which 1405 are in operation. Of these, only 503 comply with their Special Conditions of Discharge (CNA,

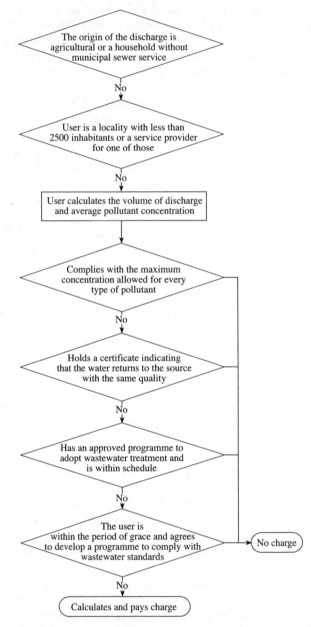

Figure 4.2   Process in the 2002 Federal Law of Charges

2002b). Only 13 per cent of the water used in industrial processes is treated afterwards (CNA, 2002a)

Table 4.5 Main industries generating wastewater discharges (m³/s)

Industry	Discharges
Sugar	45.6
Chemical	13.4
Petrochemical	7.0
Paper and cellulose	4.5
Steel	4.5
Textile	2.9
Beer	1.4
Food	1.2
The rest of the sector	2.6

Source: CNA (2002a).

The most common type of treatment for industrial discharges is secondary treatment. In recent years the treatment of industrial discharges, conditioned by the huge initial investments required, has been postponed by some industries given their lack of available funds.

The sugar industry is a special case, because of the level of pollutants it discharges and because the whole process is labour intensive, employing hundreds of thousands of workers in some of the poorest regions of the country. It is also one of the industries that have had more problems changing; they have obsolete equipment that generates higher water demands and a greater pollution load.

The type of pollutants generated by chemicals requires a more advanced treatment than stipulated in the current regulatory framework. In fact only 17 per cent of the wastewater generated from chemicals is treated. As for the discharges from petrochemicals, 90 per cent is treated; however, there is a problem related to the presence of toxic and heavy metals (CNA-Semarnap, 2000).

Mexican Official Standards (NOM)
At the core of environmental water management in Mexico are instruments for direct control, such as the Mexican Official Standards (Normas Oficiales Mexicanas, or NOMs). These are defined as a set of

physical, chemical and biological parameters and their maximum permissible levels in wastewater discharges. In particular cases, the law also recognizes Special Conditions of Discharge (CPD).

The use of the Mexican Official Standards simplified the system in place. The approach introduced in 1997 allowed for the reformulation of 43 official standards – mostly for specific industry sectors – into only two general ones: the NOM-001-ECOL-1996, specifying the permissible discharges directly into federal water bodies, and the NOM-002-ECOL-1996, for discharges into public sewers.

In September 1998 a new NOM, the NOM-003-ECOL-1997, was published to establish the standard for the use of wastewater for public services. In 2002 the NOM-004-ECOL-2002 was published to control sludge and bio-solids originating from wastewater treatment plants and sewerage systems.

The NOM-001-ECOL-1996 allows for a period of grace. The main purpose of this approach was to provide incentives for polluters to adopt new practices, processes and technologies for the reduction of their

Table 4.6   Original deadlines to comply with the NOM-001-ECOL-1996

Compliance as of:	Range of the population	Number of population centres (1990 INEGI census)
1 January 2000	Greater than 50000 inhabitants	169
1 January 2005	Between 20001 and 50000 inhabitants	181
1 January 2010	Between 2501 and 20000 inhabitants	2266

Non-municipal discharges (pollution load)

Compliance as of:	BOD (tons/day)	SST (tons/day)
1 January 2000	Greater than 3.0	Greater than 3.0
1 January 2005	From 1.2 to 3.0	From 1.2 to 3.0
1 January 2010	Less than 1.2	Less than 1.2

polluting emissions in a credible time frame. The change was driven, among other things, by the fact that the government perceived a large default among users associated with the economic crisis that had struck the country. It was therefore considered most appropriate to implement a gradual approach. And in principle this approach would give more time for authorities to improve their monitoring capacity.

The deadlines for complying with the maximum permissible levels for discharges to national waters were set for the years 2000, 2005 and 2010, according to population size in the case of municipalities and the range of biochemical oxygen demand (and/or total suspended solids) for non-municipal discharges (see Table 4.6). Thus water charges will also be scheduled accordingly. Once all deadlines are met in 2010, charges will be fully applied.

4.4 IMPLEMENTATION PHASE

This section presents the evolution of the implementation of water charges in Mexico, identifying its successes and failures.

4.4.1 Water Charges

The correct application of water charges could be very positive. Unfortunately the system still does not reflect the real costs of water and is highly subsidized. The federal government has absorbed the necessary infrastructure costs to provide water services, and revenues have made little contribution to financing.

Traditionally there have been important cross-subsidies among sectors. Municipal and agricultural water is subsidized at the expense of industry. In practice industry is a captive payer. Water for agricultural use – almost 80 per cent of the total – is free of charge.

Direct subsidies are often provided to municipalities, water system operators and irrigation works to finance infrastructure investments. The net effect is to lower the cost of water for all final consumers. This can trigger a vicious circle of lower unit prices, encouraging users to consume higher quantities than they otherwise would at full-cost pricing, inducing service providers to increase supply even further, and thus leading to overstressed and poorly managed infrastructure.

Clearly there is an important distortion in the system that necessarily reflects in a poor performance of the use of water charges. The subsidization of water services is considered as only a transitional measure in order to make moves towards full-cost pricing easier for consumers, or as well-targeted welfare measures for disadvantaged members of society. The problem is that in Mexico this fundamentally transitional measure has been a very long-term one. The situation is getting more difficult to sustain, as federal investments are being overwhelmed by demand.

Wastewater charges

As for the wastewater charge, the situation is even more complex. It is clear that the framework emphasizes compliance with the standards and traditionally has contemplated a set of economic incentives in favour of users that adopt processes for better water quality than established in the standards. Nevertheless the objective of the charges has not been fully achieved.

The increase in revenue, however, is noticeable, although with a marked historical slowdown that corresponds to the years of the economic crisis. The effect of the policy in terms of inducing the pattern of resource use is also perceptible, at least in the case of the wastewater sector. The percentage of treated water increased from 10 per cent in 1996 to 23 per cent in the year 2000. Still, that is not enough. Despite the efforts made, wastewater treatment is very limited. In the industrial sector the figure is considerably smaller, with only 13 per cent of the wastewater being treated (CNA, 2002a).

Wastewater charges have not been effective enough to alter conduct to reverse the long-term trend, nor have they reached all polluters. A crude example is the basin of the Lerma river, one of the main rivers of Mexico and an important water source for Mexico City and Guadalajara, among many others. In 1990, the river had a concentration of COD of 13.5-mg O_2/l. By 1998, the concentration was 92.33 mg O_2/l. Every other pollutant at least doubled. Dissolved solids tripled from 1994 to 1998 and oxygen in water diminished from 5.76 mg O_2/l in 1996 to 0.7 mg O_2/l (Semarnap, 1999).

Table 4.7 provides a summary of the wastewater produced in the country and the level of treatment it receives. There is a volume of around 420 m³/s of wastewater only from municipal and industrial uses. Assuming a conservative operating cost of US$0.10 per cubic metre that receives secondary treatment, the total treatment cost could be estimated at around US$1325 million a year. That is without taking into account maintenance costs or infrastructure investment.

Table 4.7   Wastewater produced in Mexico by municipal and industrial discharges

	Municipal discharges	Industrial discharges
Water used	263 m³/s	193 m³/s
Wastewater produced	250 m³/s	170 m³/s
Wastewater discharged in sewer	200 m³/s	NA
Water treatment installed capacity	75.9 m³/s	NA
Water receiving some kind of treatment	45.9 m³/s	25.3 m³/s
COD generated	1.94 million metric tons/year	6.16 million metric tons/year
COD collected in sewer	1.56 million metric tons/year	NA
COD removed by treatment systems	0.36 million metric tons/year	0.80 million metric tons/year
Percentage of wastewater treatment plants complying with the NOM 001	NA	36%

Note: NA = not available.

Source: CNA (2002a).

Figure 4.3 shows the evolution in the revenue from the use of water receiving bodies. The amount collected represents a very low percentage of CNA's revenues and a minimum amount for what is needed to maintain the operation of treatment plants, even at a cost of US$0.001 per cubic metre.

Figure 4.3 also depicts the volatility of the charge collected. When analysing these figures in US terms, it is important to consider that Mexico has faced an important devaluation in the period analysed and this factor should be taken into account. Nonetheless the increase in revenue shown until 1996 can be partly explained by the measures taken by the CNA to increase collection at the time, such as an enlargement of the register of charge payers and the publicity directed towards users. The reduction in collection presented in recent years is mainly

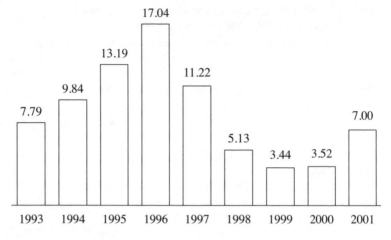

Source: Based on Presidencia de la República (2002) and INEGI (2002).

Figure 4.3   Federal revenue from wastewater charges (millions of $US)

explained by the extension given to polluters for the construction of their abatement facilities. In general the amount collected from the wastewater charges represents a very low percentage of CNA's revenues and a small amount compared with water use charges (less than 1 per cent estimated in 2002). Again it is clear that the objective of environmental authorities with the introduction of the wastewater changes was to force payers on a reasonable timetable to comply with regulation charges rather than increasing the revenue level.

The greatest problem faced by the system in recent years is that municipalities and most industries still do not pay for their discharges. As expected this default increased with the economic crisis in the late 1990s.

A lot has been done to give incentives to treat water. Yet the majority of polluters still do not comply with standards. By the end of the deadline for compliance with the NOM-001, only a few of the 139 population centres with more than 50000 inhabitants in the country met the conditions. By mid-2002, an additional small number of population centres improved significantly their water treatment processes, though to a lesser extent than required.

As for industrial discharges, the percentage of compliance was slightly better, but the figure has not been disclosed. From personal communications, it seems that some progress has been made but many problems still remain in certain sectors, such as the sugar industry, the major discharge generator, which basically did not comply with the standards.

The Water Program 2001–2006 acknowledges a low capacity to verify compliance with the standards. This programme, however, proposes ambitious monitoring goals to be achieved during the period.

Authorities recognize that there is not enough equipment or personnel to supervise the whole system. Despite being a very large agency in terms of number of employees (around 17 170), the number of staff members currently working on inspection and metering activities is very limited (less than 200 people). This situation is aggravated by the fact that even when non-compliance is detected, the effective sanction requires a complex and long administrative process.

In the past, monitoring visits were scheduled randomly, but they are now oriented to inspect the main polluters at least once in a three-year period. In total, some 3000 polluting firms are targeted for monitoring. The majority of the firms inspected in 2002 did not comply with the standards, but only a few of them were finally sanctioned. Hence, the incentives to comply are minimal.

Three presidential decrees were published in October 1995 to cancel the accumulated charges, given the large default among users and the economic situation of the country. The objective was to erase the historic debt of users that regularize their status. Certainly the burden of the debt was a serious threat to the financial health of most operators. The decrees were extended the following year, reinforcing a wrong signal of lack of commitment.

On 21 December 2001, President Fox issued two new decrees exempting payments from wastewater charges and for the use of federal water for cities with more than 20 000 inhabitants (Secretaría de Gobernación, 2001). The decree also recognized that the majority of municipalities, operating agencies or state commissions that presented their action plan 'did not comply with it because of the lack of financial resources'. The debt of the 2400 municipalities in the country, including payments due from five years previously, added to 65 000 million pesos (equivalent to US$6.5 billion at a rate of 10 pesos per dollar).

4.4.2 Financing Schemes of Water-related Infrastructure

A great portion of the funding of the water sector still comes from the

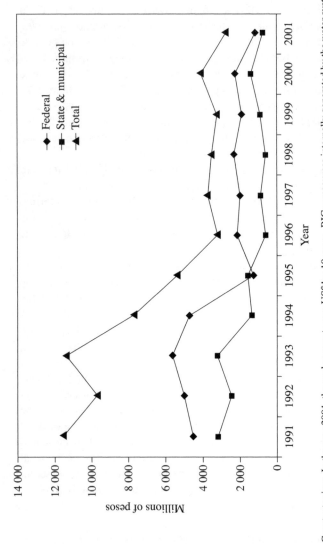

Notes: Constant prices. In the year 2001, the exchange rate was US$1 = 10 pesos. RIG = resources internally generated by the water system operators. Origin of the resources: 2001 = 100.

Source: CNA.

Figure 4.4 Public investment and financing sources for water supply, sewerage and sewerage treatment, 1991–2001

federal government budget. Water charges are not enough to cover costs, much less to finance the operation of the needed infrastructure.

As can be seen in Figure 4.4, federal investments in the sector have been falling in real terms. More and more, the transfer of federal resources amounts to response to emergencies rather than to efforts to meet the objective of increasing efficiency. These transfers come from different sources, funds and programmes. It is common to see that these transfers are not accompanied by ways to identify the impact of such investments, as they do not include clear and measurable indicators and a well-defined follow-up system (Kemper and Alvarado, 2001).

It is estimated – under the conservative scenario of the 2001–2006 National Water Program – that between the year 2000 and 2025, the investments needed to meet the minimal goals established in the water and sanitation sector will be in the order of US$41 billion. This represents an average of US$1.6 billion per year, mostly related to sewerage and wastewater treatment and not taking into account operation and maintenance costs.

In recent declarations President Fox updated these figures, asserting that Mexico needs to invest at least US$2.2 billion a year to meet the goals (Fox, 2002). This is between five and six times the CNA's budget and the revenues generated internally by the operating agencies, and including international loans, and it is not clear where these financial resources will be found. Until now, charge revenues have made little contribution to financing and the sector has depended for its development on the federal budget.

A very relevant federal programme, the Infrastructure Investment Fund (FINFRA), was created in September 1995. This programme is currently known as the programme for the modernization of water system operators (Programa para la Modernización de Organismos Operadores). PROMAGUA is the backbone of the new effort to promote investment opportunities and finance infrastructure projects. It involves federal resources to encourage higher private sector participation in the development of basic infrastructure including water and sanitation services. In particular PROMAGUA is directed towards water infrastructure in cities of more than 50000 inhabitants.

Private participation has been particularly scarce and unstable. Only 39 per cent of contracts in water and sanitation signed since 1991 were still in effect ten years later. With respect to wastewater treatment, the situation is even worse. Out of the 41 contracts signed between 1991 and 1999, only 11 are still in operation (Morales Reyes, 1999). The

Water Program 2001–2006 recognizes that wastewater treatment investments represent an excessive risk for private investors, because of the difficulty in recovering their investment with the actual tariffs and collection rates.

4.5   CONCLUSIONS AND RECOMMENDATIONS

In Mexico, economic instruments for water management have been in use since the 1980s but their effectiveness is still difficult to evaluate because they have not been fully functional.

There are some specific and evident problems that need to be solved in order for the system to start to give better results. Some recommendations to be implemented in the short term include the following.

Improve Monitoring

It is hard to determine whether this problem has a financial root or a political one, but the system has been functioning in a vicious circle of low prices and low investment. The circle has been aggravated by a lack of enforcement of the regulations by the authority.

Eliminate Most Exemptions

Water charges have been subordinated to control mechanisms and policy interests, and as a result they have been corrupted as true economic instruments. The most notable example is the way wastewater charges are determined by a set of pollution control standards. In the future, when wastewater charge exemptions meet their final deadlines for full compliance with wastewater discharge standards (NOMs), the real potential of wastewater charges will be tested.

Consider Reducing Subsidies

Some subsidies, especially cross-subsidies, can be unjustifiably big and often have nothing to do with the nature of the charges. These could be reduced.

Charges should be based on real costs. For example wastewater charges should charge for the pollution added, and therefore, they must take into account the pollutants of the water intake and not only the volume of pollutants discharged.

Improve Local Capacity and Interest in Water Management

The CNA has made important improvements in the construction of a unified national water system. Yet it has not been so successful in its decentralizing efforts. Participatory organisms, such as River Basin Councils, have played a minor role in water management so far. Also there have been weak incentives for states and municipalities to comply with the federal law and thereby to reinforce the federal system and apply control mechanisms inside their territories and jurisdictions.

NOTE

1. This chapter was part of a series of papers commissioned by the Inter-American Development Bank for the Environmental Policy Dialogue and the opinions expressed in this chapter are solely those of the authors and do not necessarily reflect the position of the IADB. Exchange rate used was US$1 = 10 pesos, average rate in 2001.

REFERENCES

Brañes, Raúl (2000), *Manual de derecho ambiental mexicano*, Mexico: FCE.

CNA – Comisión Nacional del Agua (2002a), *Compendio Básico del Agua en México*, Mexico.

CNA – Comisión Nacional del Agua (2002b), *Situación del subsector de agua potable, alcantarillado y saneamiento a diciembre de 2001*, Mexico.

Fox, Vicente (2002), Speech at the presentation of the evaluation results of Promagua, Los Pinos, Mexico, 17 September.

Giner de los Ríos, F. (1997), *Los instrumentos económicos y la regulación ambiental en México en Economía Ambiental: Lecciones de América Latina*, Mexico: Instituto Nacional de Ecología, Primera edición, December.

Guerrero, H. (2002), 'Irrigation water pricing: the case of Mexico', paper presented at the International Conference 'Irrigation Water Pricing: Micro and Macro Considerations', 15–17 June, Agadir, Morocco.

Guerrero, H. and C.W. Howe (2000), 'Water pricing in Mexico: principles and reality', European Association of Environmental and Resources Economists (EAERE), Tenth Annual Conference, University of Crete, Department of Economics, 29 June–2 July.

INEGI (2002), *Estadísticas económicas de mediano plazo*, Mexico. http://www.inegi.gob.mx/.

Kemper, K. and O. Alvarado (2001), 'Water', in Marcelo M. Giugale, Olivier Lafourcade and Vinh H. Nguyen (eds), *Mexico: A Comprehensive Development Agenda for the New Era*, Washington, DC: The World Bank.

Morales Reyes, Javier I. (1999), 'Situación y Propuestas para el tratamiento de aguas residuales y participación privada en México', *Federalismo y Desarrollo*, No. 65 January–March, Mexico.

Presidencia de la República (2002), *2o Informe de Gobierno*, September, Estados Unidos Mexicanos.

Roemer, A. (1996), *Derecho y economía: Políticas públicas del agua*, Mexico: FCE.

Secretaría de Gobernación (2001), 'Diario Oficial de la Federación' states: 'Decreto por el que se condonan y eximen contribuciones y accesorios en materia de derechos por uso o aprovechamiento de bienes del dominio público de la Nación como cuerpos receptores de las descargas de aguas residuales a cargo de los municipios, entidades federativas, Distrito Federal, organismos operadores o comisiones estatales o responsables directos de la prestación del servicio de agua potable, alcantarillado y tratamiento de aguas residuales', 21 December, Mexico.

Semarnap (1999), *Estadísticas del medio ambiente*, Mexico.

Semarnap-CNA (2000), *El agua en México: retos y avances*, October.

Semarnat-CNA (2001), *Programa Nacional Hidráulico 2001–2006*, Mexico.

Tecsasím-Lyonnaise des Eaux (2000), *El libro del agua. La Participación Privada en el sector*, directed by Alain Biche, March, Mexico.

5.   Country case: Brazil[1]

José Gustavo Feres and
Ronaldo Seroa da Motta

5.1   POLICY ANALYSIS PHASE[2]

This section presents an overview of water policies in Brazil. It first analyses the previous legal and institutional frameworks and then describes the development of the new water management policy under which the use of water charges in Brazil was introduced.

5.1.1   An Overview of Water Management in Brazil

The first water regulation in Brazil was the 1934 Water Code. At the time it was a very advanced regulation in determining multiple water uses, although it was mostly related to quantity aspects to accommodate agricultural, urban and energy uses. It also created the water permits issued to assign individual water volumes and introduced the distinction between federal and state rivers. Federal waters are those that flow across or along the boundaries between two or more states or a foreign country. State waters are those situated entirely within the territory of a single state. This discrimination may sound too complicated, but it is, in fact, an option to deal with water resources in a country of Brazil's size and hydrological complexity and abundance.

The code also created an institutional arrangement led by the National Department of Water and Electric Energy (DNAEE). States also followed the same centralized and energy-oriented structure.

Direct regulation on pollution issues arose during the 1970s with the process of rapid industrialization. The first Brazilian law on pollution control was created in 1977 and included the monitoring and control of water quality. Its main concern was on the physical-chemical conditions that are required to keep water suitable for drinking. The same law also established some penalties to polluters but without further specific reference on how to enforce them. This law was never fully enforced due to the political resistance within the government on penalizing state-owned sanitation companies, since domestic wastes were by themselves the main pollution source.

The current referential on water quality ambient standards is under the CONAMA Resolution No. 20 of 1986, which introduces five classes of water according to the quality needed to support types of use.[3] This resolution states that all federal waters must be classified accordingly and that state waters should also be classified following the minimum parameter values defined for federal waters. The states are responsible for the classification as well as the resulting monitoring and surveillance of their waters.

A timid version of the river basin approach was first sought in Brazil with the creation of the Special Committee of Integrated Studies of River Basins (CEEIBH) by an inter-ministerial resolution (Portaria no. 90 of 28 March 1978). This resolution stated that CEEIBH would carry on studies suggesting classification and monitoring of river basins aiming at the sound use of water resources in Brazil, considering the multiple uses and minimizing environmental impacts. Since then, executive commissions of several river basins were organized in order to undertake these studies, such as the São Francisco and Paraíba do Sul river basins.

The institutional arrangement established that the CEEIBH presidency and executive secretariat would be occupied alternately by DNAEE and the existing federal EPA (SEMA and later IBAMA) with a two-year mandate. Other members of CEEIBH are municipal administrations, sanitation companies, electric power companies, state environmental agencies and other official agencies related to water services such as irrigation.

The major achievements of CEEIBH were the undertaking of several management plans and the classification of rivers. Since the committees had no normative power and financial autonomy, most of their recommendations were rarely put in practice. Also, their members' engagement in meetings and report elaboration was reduced due to financial limitation.

However, this experience guided the new constitutional requirement on water resources control and management which institutionalized the National Water Resources Management System.[4]

5.1.2   The Development of the New Water Policy

To accomplish this constitutional requirement, CEEIBH and water management experts, with support from DNAEE and IBAMA, set up a series of meetings to discuss the issue and to prepare a version of the legislation required for the implementation of this new system. By the end of 1989, a draft bill was presented in the National Congress, according to which DNAEE would take over the previous CEEIBH arrangement with more centralized regulation prerogatives.

This draft was very much tailored for energy production concerns, already threatened by water permit disputes with irrigation and urban uses. Financial autonomy would be initially achieved through compensation payments from revenues of hydroelectric production (later approved). The version also emphasized the need for the application of water charges to finance the system and to make more efficient the water uses in River Basin Committees.

Although River Basin Committees would be responsible for water management in their own jurisdiction, they would be inserted into the governmental structure and subordinated to a federal council. That would include federal as well as state rivers. In sum, the structure would be almost another governmental level in Brazil with a high fiscal autonomy and controlled by the energy sector.

Opposition from state and municipal governments, users and environmental agencies was promptly conveyed in the National Congress. To accommodate these conflicting interests the lead in the debate of the draft bill in the National Congress was taken by the former Secretariat of Strategic Affairs directly linked to the Presidency of the Republic. With that DNAEE has lost much of its previous authority to interfere in law-making.

In 1991 the state of Ceará accepted an offer from the federal government and took control of federal dams and reservoirs built to tackle water supply shortage in the state in an attempt to decentralize drought policy in the north-east of the country. The state government then took the opportunity to redefine the Ceará water resource policy. It started by creating users' associations for each reservoir-impacted area and applying water pricing. This pricing was aimed at the need to recover the operational costs of water supply. Note that the Ceará case is typical of a drought region where scarcity is the crucial element and water storage is the main operational measure.

In 1992, the state of São Paulo approved its Water Resource Policy law that was clearly similar to the French system in terms of: (1) a compulsory creation of 14 River Basin Agencies in the state, and (2) the introduction of water charges for quantity and quality elements based in a common pricing criteria for the whole state.

Although each basin would have its own committee, water agencies would be public entities linked to the State Water Resource Council headed by the Secretariat of Water Resources. Water charge revenues would be credited to the State Water Resource Fund. Application of revenues would follow the State Water Resource Plan and the basin's investment plans would be accommodated accordingly.

At the same time, the Commission of Environmental Issues in the National Congress prepared a new draft in which the leadership was passed to the environmental sector but with the same hierarchical river basin structure and governmental centralization. At this point water charges were not clearly defined, although they were always presented as one of the major instruments to deal with water scarcity and contamination.

A series of debates took place across the country with state water institutions, water experts and users' representatives. From 1994 onwards, with the creation of a specific Secretariat of Water Resources within the MMA structure, the discussion was speeded up and a new national draft was elaborated accommodating the conflicting interests. The major changes were:

1. A clear, integrated water management approach under the leadership of the Ministry of the Environment.
2. An explicitly decentralized system with no compulsory creation of River Basin Committees in the country.
3. Reduced federal government interference and amplified autonomy to River Basin Committees.
4. The National Council would be restricted to general policy and principles would integrate federal and state river basin systems and would be composed of all governmental levels and users' representatives.

Note that this law is a national one and therefore its principles apply to both federal and state rivers as well as to underground water. However, its specific regulation of basins and instruments is only enforceable on federal waters. The option for a non-compulsory hierarchical basin system resulted from the recognition that a country the size of Brazil and with a complex river system would not be able to implement a hierarchical and compulsory system due to constraints in institutional capacity, including in human resources.

At this time, the energy sector was the major opposition force since it was very suspicious that this decentralized approach would not be able to cope with the sensitive needs of the hydroelectric system. With regard to the water charge system, the suspicions were on the side of the Ministries of Planning and Finance, that were concerned with the potential excessive fiscal importance of the water charges. While the opposition of the electric sector was defeated within the National Congress debate by the strength of the MMA, the compromise on fiscal issues was achieved by requiring federal administration of the water charge revenues.

In parallel, since 1992, state water legislation has been approved in 14 states. Some were later revised to accommodate the 1997 National Policy's principles. Most of them have the same French-oriented structure with management dominance by the state government, as adopted in the São Paulo case. This trend in the states can be understood first due to their less complex river structure and second, if not mainly, to the need to strengthen states' participation at the CNRH where conflicts of federal and state river basins will be intermediated.

Apart from the differences on the degree of decentralization regarding the role of River Basin Committees (CBH), all laws apply the same instruments, such as: (1) River Basin Management Plan (PRH), (2) issuance of Water Rights, and (3) Water Charges.

Only Ceará is already applying charges for bulk water. São Paulo already has a regulation on charges but it has not been able to apply them yet. These two cases will be further explored later in this chapter.

5.2 THE DESIGN PHASE

This section focuses on the formulation and design of water charges in Brazil within the new water policy framework.

5.2.1 Institutional Arrangements

As said before, the basis of the new federal water management system in Brazil is decentralized and not compulsory. This decentralized pattern is less present in state legislation.

It means that in principle only those federal basins that decide to constitute committees are required to comply with the law. The National

Council of Water Resources (CNRH) regulates the constitution of a committee according to specific representation criteria.

CNRH is the national water policy-making institution headed by the Ministry of the Environment, but its composition covers federal and state representatives, sectoral users (industry, energy, agriculture and so on), as well as civil society.

Enforcement of the water regulation is undertaken by the National Water Agency (ANA), an executive body with its own career staff and directors having a four-year mandate (in periods not coincident with the federal administration's). ANA is financially strong since it has a share of payments related to the hydro-energy compensation (*compensação financeira*) created by a law passed in the early 1990s. This payment is 5 per cent of the hydroelectricity gross revenue sales, to compensate the land area used for reservoir construction. It is distributed according to the size of the restricted area to municipal (major share), state and federal (smallest share) governments. This compensation revenue was diverted to ANA from the DNAEE and in 2002 represented a budget of around US$30 million dollars, a third of ANA's total budget.

ANA also became the manager of water charge revenues. Charges are determined by River Basin Committees (CBHs) following general pricing principles to be dictated by the CNRH. River Basin Agencies (ABH) collect the charges for ANA that later returns the resulting revenue to basin agencies. The law allows up to 7.5 per cent of charge revenues to be used for administrative purposes, and it is not clear about the partition of this share among CBHs and ANA. The law vaguely specifies that should be 'mostly' used in the basin itself.

As said before, ANA control of charge revenues was the political way of assuring a higher degree of control to the federal government, in political as well as fiscal terms. Not surprisingly, states, municipalities and users are very concerned how ANA will exercise its revenue distribution power.

The major factor that made possible the approval of the law was its decentralized structure and financial autonomy. As worldwide experiences confirm, the return of revenues to the contributing basins is the primary incentive for cooperation in CBHs.

Although the legal mandate of the CNRH is for planning and that of ANA is for supervision, due to the need to regulate the new water law there are jurisdiction conflicts between both institutions on policy matters. This is clear in regard to the charge system when ANA is in fact leading the debate in the CNRH on the elaboration of the charging principles, norms and schedule.

River Basin Committees are formed by users, but federal, state and municipal governments can have up to half of the seats. Each CBH has its own agency, although an agency can work for more than one CBH.

Following the French experience, the River Basin Management Plan is a five-year plan that defines the environmental objectives for water availability and quality in the basin in terms of legal standards, type of uses and water rights through the whole period. It also indicates the type, size and schedule of water-related investments and specifies their financing schemes, such as loans, public budget allocations and the water charge levels. It is prepared by the River Basin Water Agency and approved by the River Basin Committee. Therefore the PRH is the common ground for articulating objectives and means in the basin management. And, as can be seen, water charge revenues are articulated with basin's investment needs and environmental targets.

5.2.2 The Water Charge System

Although in Brazil the new federal water management system has adopted the same French principles of management by water basin, it has a difference in regard to charges. The water charge (*cobrança pelo uso da água*) in the Brazilian case can be seen as a kind of service payment that is defined within the Basin Committee. There is no predefined general charge structure as in France, and committees have the autonomy to set their own charging criteria. Again this highly decentralized approach was needed to take into account the complex river system in Brazil.

Also from the legal point of view, it was a way as to avoid classifying the water charge as a tribute (tax or duty), that would require a specific additional law to change the Tax Chapter of the National Constitution, and therefore face a much more difficult political process.

The first implementation of water charges in federal rivers started in March 2003 in the Paraíba do Sul river basin, where a simplified charge system was implemented and, according to the results, the Committee defined new pricing criteria.

Although in the mid-1990s some states approved new water laws with the use of charges, no state has been able to fully implement the law so far. Apart from institutional weakness in setting up the river basin system and creating committees and agencies over the whole territory, the main constraint is the pricing criteria to be adopted in the charge system.

All laws take the idea that water has an economic value and uses should be rationalized to allow attainment of multiple ends. Table 5.1 presents the objectives of water charges as assigned in the texts of the national and state water laws in Brazil (also indicating the law issuance year). As can be seen, they all apply charges to generate revenue to finance water-related investments in the basins as well as to attain improved water quality and availability. In sum, the proposed water charge systems have both revenue and effectiveness objectives.

They all also discriminate by type of use. This is a clear indication that water volume will be valued differently by type of user, and consequently cross-subsidies among users will be allowed. Some states go further in discrimination, clearly saying that the socio-economic conditions of the user can affect its charge level as an indication that equity issues will be accounted for. Others wish to use charges to promote regional objectives. This may also imply that charge levels will vary to protect activities that represent regional objectives. In most cases they accept cross-subsidy among basins allowing for the application of charge revenue outside the basin. Only one state, Mato Grosso, indicates a criterion related to charges being used for changing spatial distribution of economic activities.

The National Law does not specify in its text a charge criterion related to the user's socio-economic conditions and it only accepts charge exemptions for those uses with very low impact for which monitoring can be too costly. No law, however, has indicated how these objectives can be combined in common pricing criteria. Instead they all have some indications of criteria that water charges must follow.

In sum, law texts only reveal the macro objectives of water charges. Since all indicate investment financing, environmental aims and discrimination among users, there will be a wide range of possibilities to use charge levels in any direction just by assigning more weight to one criterion in detriment to others. So the debate on regulation of the water charge system will reflect these conflicts.

5.3   IMPLEMENTATION PHASE

There are only three experiences of water charge regulation in Brazil. A simple cost-recovery approach adopted in the state of Ceará has been in

Table 5.1 Water charge criteria in Brazilian water laws

States (issuance year)	Revenue generation to finance basin's investments	Improve environmental quality	Type of use	Socio-economic conditions of the user	Regional economic objectives	Inter-basin revenue application	Modify space occupation
Alagoas (1997)	X	X	X	X	X	X	
Bahia (1995)	X	X	X	X	X	X	
Ceará (1992)	X	X	X				
Distrito Federal (1993)	X	X	X				
Espírito Santo (1998)	X	X	X	X		X	
Goiás (1997)	X	X	X				
Maranhão (1997)	X	X	X	X	X		
Mato Grosso (1997)	X	X	X	X	X	X	X
Minas Gerais (1999)	X	X	X				
Pará (2001)	X	X	X		X		
Paraíba (1996)	X	X	X			X	
Paraná (1999)	X	X	X				
Pernambuco (1997)	X	X	X		X	X	
Piauí (2000)	X	X	X				
Rio de Janeiro (1999)	X	X	X				
Rio Grande do Norte (1996)	X	X	X	X			
Rio Grande do Sul (2000)	X	X	X			X	
Santa Catarina (1994)	X	X	X			X	
São Paulo (2000)	X	X	X				
Sergipe (1998)	X	X	X			X	
National (1997)	X	X	X		X		

place since 1995. In São Paulo there is already a draft bill being discussed. The third is the experience in the Paraíba do Sul River Basin Committee, started in March 2003. All three cases are described next,[5] in particular the Paraíba do Sul river case which is the most complete case of charge regulation in implementation in Brazil.

5.3.1   Ceará Case

In the State of Ceará, water resource management is a special case in Brazil and its state water law is simpler than those of the other states. It did not adopt the full river basin approach and water agency arrangement. Water supply in the state depends on reservoirs and dams from which water is distributed along natural and artificial channels. For a long time a bulk water tariff system has been in place for operational cost-recovering purposes. When in the early 1990s the state took control of federal reservoirs, almost 60 per cent of the supply, the state government decided to revise its water management. Planning and full pricing were the main instruments.

The pricing criteria adopted were simple and charges vary with transport costs (from reservoirs to consumption point), user's ability to pay, and degree of supply assurance. The administration is undertaken by a state water supply company (COGERH) and there is a State Water Council with governmental and non-governmental representatives that dictate water policy where COGERH acts as an executive body.

Until recently, each reservoir area had its water user association that was encouraged and assisted by COGERH to discuss and plan water supply and distribution criteria, although state intervention through the State Water Council and COGERH was dominant. With the new National Water Policy the state also introduced the river basin management approach and increased autonomy of committees.

In the beginning, only industrial and domestic users (through sanitation companies) were charged, but currently agriculture and aquaculture are charged for bulk water use as well, as shown in Table 5.2. As can be seen in the table, prices for industrial users are almost 28 times higher than those for domestic users, who in turn pay charges as much as six times higher than rural users.

In the first years of this decade, charges revenues have reached the level of about US$2 million per year in order to cover the COGERH's operational costs. Note that the COGERH pricing criteria do not capture investment costs. That is, they do not apply long-term marginal cost

Table 5.2   2001 bulk water pricing structure in Ceará (US$/1000 m³)

Users	Charge Levels
Industrial	327.66
Domestic	5.53–11.91
Irrigation and aquaculture	0.43–2.13

Note: 2001 average exchange rate of R$2.35 per US$.

Source: COGERH.

pricing. Funding of investments is taken from the state general budget. Ceará is one of the poorest states in Brazil. Therefore there is not much room to increase water prices for low-income domestic users and farmers. This distributive criterion makes the COGERH system politically viable, although the company is willing to allow prices to rise gradually (though they may not reach full cost recovery), as an 'educational' indicator of water scarcity to adjust demand in the long term.

However, with the recovering of operational costs alone, COGERH was able to guarantee a 99 per cent level of supply assurance for the industrial sector, 95 per cent for urban consumers and 90 per cent for agriculture.

So the case of Ceará is very impressive. However, one should take into account the particular characteristics of this state's water supply and demand before advocating its replication in other regions of Brazil. Water scarcity in this state is very high since there are no perennial rivers in the region, therefore water management is of paramount importance to overcoming seasons of drought. Also there is no major federal river in Ceará, and consequently the state government is free to implement its own policies, without waiting for federal regulation.

Ceará has also not attempted to introduce pollution components into its water pricing structure, thereby avoiding a rather complicated matter. This simplifies the problem, as the implementation of a bulk water pricing structure based on quantity is similar to those of utilities such as energy.

The Ceará case, rather than a generalization, should be seen as a particular solution to a particular case. Although a cost-recovery system has not been fully applied, the main message is that water pricing is feasible even in poor regions, and can play a decisive role in better water management and increased user participation.

5.3.2 São Paulo Case

São Paulo state is the most advanced in charge regulation in Brazil. The state first elaborated a proposal on water charges to regulate its 1991 water law. In October 1997, the State Water Resource Council (CRH) of São Paulo started discussing this proposal[6] (CRH, 1997) to define specific water charges for all types of use, including irrigation, recreation and navigation. The proposal advocated setting charges based on a basic unit price (*PUB*) from which the total charge bill is calculated. As a ceiling for the total charge bill there is a maximum unit price (*PUM*) and also a total average charge cost (*CMR*).

The proposed pricing methodology is straightforward and very similar to the one applied in the French system (see Chapter 3). The total amount charged to a user for use j in basin i ($CT_{j,i}$) is calculated by multiplying PUB_j by the quantity of water volume consumption (intake minus returned volume) and volume used for dilution of pollutants (quantity to dilute pollutants according to environmental regulation) expressed as ($Q_{j,i}$) and by use and basin related coefficients indicated in ($X_{j,i}$), so that:

$$CT_{j,i} = Q_{j,i}\, PUB_j\, X_{j,i}$$

where $X_{j,i}$ are use and basin characteristics (type and location of use, basin water availability and quality, and so on) that will be decided by Basin Committees provided that the $PUB_j\, X_{j,i}$ product does not exceed PUM_j. In addition to that, the sum of all user's $CT_{j,i}$ cannot exceed a proportion of the total water bill (that includes water supply and sanitation services) given by CMR.

The proposal also suggested a gradual introduction of $X_{j,i}$ factors, starting from types of use, moving then to basin and use characteristics including eventually more complex considerations, such as, peak period and seasonal effects.

Higher prices are suggested for industry, medium prices for urban use, and lower prices for irrigation. For water quality, irrigation is charged more than urban use. Rather than being based on (economically) efficient pricing, this overall approach seems to be based on revenue-raising objectives. Regarding classes of rivers, the higher the environmental quality, the greater the coefficient value. As in France, a higher price is adopted to induce greater control where the class of rivers is most sensitive.

The study estimated annual revenues from withdrawal charges and organic/solid suspended pollution charges to be around US$500 million, with approximately 50 per cent from urban consumption, 30 per cent from irrigation and 20 per cent from industry. However, this estimate assumes that price elasticity is zero, which is generally not the case. In reality it is likely that users will adjust their demand for water once they are faced with paying higher water charges, thereby diminishing the actual amount of revenues collected.

The proposal was fully discussed and later in 2000, the state government sent a Draft Bill to the state assembly (PL 676/2000) setting specific principles for water charges. The bill says that charges may vary according with the water source (superficial or underground); type, location and effective volume of use; conditions of water quality, availability and regularization in the basin; seasonal effects; and conservation measures.

PUM value was indexed as a fraction of 0.001078 of the UFESP that is the state index reference unit to correct inflationary effects on tribute payments. This fraction would represent in 2000 a value of R$0.01 per cubic metre of water (or US$0.006/m^3). The bill indicates that urban and industrial uses would be first charged and other uses (including agricultural) only after four years.

The bill has not yet been approved in the State Assembly. The debate so far has not succeeded in overcoming the controversies on water charges in the state regarding institutional arrangements, universality of uses being charged, autonomy of Basin Committees and charge revenue allocation criteria.

5.3.3 The Paraíba do Sul River Case[7]

The controversies on water charging pointed out in the São Paulo case are likely to occur within River Basin Committees whenever they discuss the application of charges within their boundaries.

Therefore, currently, the CNRH is working on a resolution on general principles for water charges. The CNRH debate counts with the participation of all governmental levels, user representatives and academia, in order to achieve consensus on all-important issues regarding water charging that has been blocking its implementation, such as those refereed in the São Paulo case. The main aim of the resolution is to agree on general principles that can be implemented as a departure point for states and Basin Committees to further elaborate their charging specific criteria.

However, the main important initiative is the implementation of water charges on the federal basin of the Paraíba do Sul river (RPB). This is an experimental case that has been promoted within the Basin Committee, CEIVAP, which has been created by a joint initiative of all boundary states of the basin.[8]

This experience has been conducted with the assistance of ANA, and the CNRH has passed a resolution setting its experimental status and, consequently, allowing that charging criteria could be defined without the definition of the national formal guidelines. As will be seen, this experience will address the most important issues on water charges in Brazil.

5.3.3.1 Environmental and economic profiles

The basin is located in the south-east region. It has a drainage area of about 55 400 km^2, distributed across the states of Minas Gerais (20 900 km^2), Rio de Janeiro (21 000 km^2) and São Paulo (13 500 km^2). About 5.6 million people live within the basin, distributed among large cities and smaller rural municipalities. The basin represents 0.7 per cent of the country's surface but, despite its modest size, it is important due to its geographical situation. The valley of the main river, which connects the two most important Brazilian metropolitan areas – Rio de Janeiro and São Paulo – flows across important urban and industrial centres accounting for about 10 per cent of the country's GDP.

Water pollution is identified as the main problem of the basin, primarily due to industrial and domestic effluents. This situation can be mostly attributed to discrepancies between the socio-economic development of the region and the insufficient measures to preserve environmental quality.

The rapid demographic growth experienced by the majority of basin urban areas was not accompanied by adequate planning and sanitation measures, resulting in the indiscriminate occupation of river banks and the lack of sanitation infrastructure. Few municipalities are equipped with wastewater treatment plants, while the vast majority disposes of their untreated sewage directly into the main river or its tributaries. According to the Paraíba do Sul Water Resources Plan, 69.1 per cent of households in the urban areas are connected to municipal sewage networks, with only 12.3 per cent of collected domestic wastewater treated before its release in the water bodies. It is estimated that domestic effluents are responsible for a BOD discharge of 240 t/day in the river basin.

The same trend can be observed in the industrial activities, whose development was not accompanied by adequate measures for industrial pollution control. The estimated daily BOD discharge related to industrial activities is about 40 t/day.

Table 5.3 sheds some light on the magnitude of the problem of water quality degradation in the Paraíba do Sul river basin. As can be seen, water quality parameters measured by monitoring stations indicate the high percentage of violations of the readings with respect to the mandatory quality standards established by the river class. The figures concerning phosphates, coliforms and BOD point out the excessive level of organic pollution. The significant presence of highly toxic substances like aluminum and phenol highlights the important role played by industrial pollution in the river basin.

Table 5.3   Selected critical water quality parameters in the Paraíba do Sul river basin

Parameter	Average violations* (%)
Aluminum	98.9
Phosphates	90.3
Phenol	34.4
Faecal coliforms	77.8
BOD	11.8

Note: *Percentage of readings which violates the parameter standards set according to water bodies classification defined by CONAMA's resolution.

Source: Paraíba do Sul Water Resources Plan (2002).

More than drawing attention to the water quality degradation, Table 5.3 also illustrates the failure of water pollution control in the basin. The malfunctioning of the control mechanisms can be explained by the lack of monitoring and control capabilities from the state environmental agencies in charge of water pollution control. Environmental agencies suffer from insufficient coercion instruments, human and especially financial resources, preventing them from enforcing the water regulation measures. Under these circumstances water users have been given no incentives to engage in water pollution control activities and other socially rational water use practices.

Given the critical situation of water quality and the importance of the river's geographical position, the federal government decided to define as a priority the implementation of the new water management approach in the Paraíba do Sul river basin. The reorientation towards a decentralized and participatory framework started in 1996 with the creation of the Paraíba do Sul River Basin Committee (CEIVAP). However, the establishment of CEIVAP was not followed by further implementation measures, setting the reform process back. This can be mainly attributed to pressures exercised by special interest groups, in particular the electricity sector, represented by the influential National Power Agency (ANEEL), along with some industrial sectors.

The creation of the National Water Agency (ANA) in July 2000 gave a new impulse to the reforms. ANA's commitment to the implementation of the new water management system has managed to offset the pressures exerted by interest groups, keeping the participatory principle of the new water management system. Indeed ANA has played a key role in the significant progress made towards the implementation of the new management system in the Paraíba do Sul river basin, contributing to CEIVAP's institutional strengthening and assisting in the issuance of water permits and charges.

Since 2000, negotiations about the methodology to apply water charges have proceeded according to the participatory principle. The first resolution of the water charge proposal, concerning domestic and industrial water users, was voted by CEIVAP in March 2001, and modified in November. The water resources plan was finished by July 2002, while the basin's Water Agency was officially recognized by the CNRH in October. Finally, in November 2002, CEIVAP approved the water charge methodology for hydropower generation plants, fish farms and agricultural activities.

While the Paraíba do Sul river flows across state boundaries, which places it within the federal domain, important sub-basins are entirely located within state territories and so are state waters. Management of federal and state waters requires a great deal of effort in terms of institutional coordination and pricing criteria harmonization if one intends to design and implement water charges to the whole river basin in an integrated approach.

As it will be seen later, the institutional arrangements to establish water charges that fully cover the river basin are still not complete. Federal and state administrations in the Paraíba do Sul river basin area are at different stages concerning the implementation of the water management system. For example at the federal level the Paraíba do Sul

River Basin Committee (CEIVAP) and its respective Water Agency were set up, and they are already implementing charge regulations, whereas states are less developed and still facing problems in implementing charge systems.

5.3.3.2 The water charge system

This situation motivated CEIVAP, jointly with ANA, to propose a water charge restricted to federal waters. This proposal presents a pioneering aspect, since the Paraíba do Sul water charge represents the first application of water charges, along the principles of the new water law, to a federal river.

In the definition of the water charge methodology, CEIVAP adopted simple rules both in conceptual and operational terms. This simplicity is intended to get users familiar with the water charge system, and to learn more about the way their behaviour can be modified. The option for simplicity also makes possible the implementation of the water charge over a short-term horizon, since sophisticated methodologies would require data about water quantity and quality aspects which are not currently available.

Generally speaking, the guidelines for the methodology embody the following principles:

- Simplicity. Conceptual and operational simplicity, as discussed above, were the main guidelines in defining the water charge methodology. The charge mechanism was designed to be based on directly measurable parameters in order to allow clear understanding by the users;
- Acceptability. Acceptability by the users is a fundamental requirement in order to legitimate the water charge mechanism. The participatory approach in the CEIVAP, which is responsible for the definition of the water price methodology, facilitates this task.
- Signalling. Water charges are supposed to act as signals about the economic value of water resources and the importance of sustainable use, in terms of both quantity (withdrawal and consumption) and quality (effluent dilution).
- Minimization of economic impacts. The signals, however, must not be so strong as to jeopardize acceptability. Therefore the pricing criteria were defined in order to minimize the economic impacts on users in terms of cost increases. So far this has been accomplished by adopting low values for the water charges.

One can easily notice the contradictory nature of such guidelines. In particular, the question of acceptability and minimization of economic impacts are clearly at odds with the signalling role of water charges. If charges are set at high levels so as to reinforce their signalling role, inducing water users to undertake water-saving and pollution abatement investments, the economic impact on users may be high enough to put at risk the acceptability of the charge. On the other hand, low charge levels with minor economic impact on users' cost may ease the acceptability, but at the same time can fail to give incentives to the agents to adopt sustainable water use practices. As will be later discussed, water charges in the Paraíba do Sul river basin have ended up closer to the acceptability/economic principles than the signalling one.

In March 2001, CEIVAP approved a first draft of the water charge proposal. According to this draft, only the largest industrial and municipal users of federal waters would be charged starting in 2002. The largest industrial users were defined to be the 40 most important water polluters located in the states of São Paulo, Rio de Janeiro and Minas Gerais. Concerning municipal water supply and sanitation service companies, firms serving cities with a population bigger than 10 000 inhabitants were considered large users.

Nevertheless detailed discussion and negotiation about methodology and pricing criteria effectively started after the approval of this draft. During these discussions, the gradual implementation of federal water charges was abandoned in favour of the application of charges to all water permit-holders. The proposal of a methodology for the other relevant water users in the Paraíba do Sul river basin – agriculture, hydropower generation, mineral extraction and inter-basin transfers – was defined as a necessary preliminary condition in order to implement water charges in the basin. Finally CEIVAP in December 2001 approved the enlarging of the universe of payers to all permit-holders. Only 'non-significant'[9] users, withdrawing less than 1 l/s, were exempted from paying water charges.

The transitory nature of the water charge mechanism must be noted. It is valid only for three years after the effective start of water charge implementation. Several reasons contribute to the definition of this short transition period. Firstly, the transitory character can be credited to the simplicity of the selected mechanisms, which should be replaced by more sophisticated ones as users become familiar with this policy instrument. Secondly, it is supposed that during this period states will have implemented their own water management systems and, in particular, will have approved water charge methodologies for the state waters of the Paraíba do Sul river basin. This will allow the interaction between state and federal water charges that should facilitate the application of these water charges in an integrated framework. Finally, the transitory character is due to legal factors: the National Water Resource Council (CNRH) allowed this experience under exceptional grounds, since water charges are not completely regulated at the federal level.

In the following sections we review the approved and proposed water charge mechanisms for the different sectors. From the discussion, it can clearly be seen that this first approach of water charge criteria was almost exclusively driven by revenue-raising purposes, with little importance given to the application of water charges as a means to promote efficient water use. In short, the signalling aspect of water charges was ignored.

a)   Industrial and municipal water supply and sanitation sectors
The methodology and pricing criteria concerning industrial and municipal water supply and sanitation services were approved by CEIVAP in December 2001, and ratified by the CNRH in March 2002. The implementation of the water charges was supposed to start later in 2002. Due to the delay in the issuance of water use permits,[10] a requisite to the calculation of the charge value, implementation was postponed to 2003. We remark again that this transitory methodology is intended to apply for three years from its implementation and just concerns federal waters.

The approved formula for industrial and domestic users follow closely the one adopted in the French system.

Total monthly water charge is then given by:

$$TWC = Q_w \times [\ K_0 + K_1 + (1-K_1) \times (1-K_2K_3)]\ \times PUP$$

where
Q_w = monthly withdrawal use permit (m³/month)
K_0 = withdrawal use unit price multiplier, defined by CEIVAP (less than 1.0)
K_1 = consumptive use coefficient (that is, proportion of withdrawn water that is not returned to water bodies), which varies according to the user's sector of activity
K_2 = percentage coverage of effluent treatment by the user
K_3 = efficiency level in terms of BOD reduction, which varies

according to the pollution abatement process adopted by the user

PUP = public unit price (R\$/m^3) corresponding to charges related to withdrawal, consumption and effluent dilution, defined by CEIVAP

It should be noted that the charge is calculated using the quantity assigned by the withdrawal use permit, and not the actual volume of withdrawn water. Since the transitory permits (for the next three years) are issued in a declaratory way, where users are assigned permits based on their declared use of water resources, the payers have an incentive to underdeclare their use. This feature calls for a high public monitoring and enforcement effort in order to avoid revenue losses.

The formula can be rewritten in order to identify the three types of water uses that are subject to water charges, namely: withdrawal, consumption and effluent dilution, as follows:

$$C = \underbrace{Q_w \times K_0 \times PUP}_{\textit{WITHDRAWAL}} + \underbrace{Q_w \times K_1 \times PUP}_{\textit{CONSUMPTION}} + \underbrace{Q_w \times (1-K_1)(1-K_2K_3)PUP}_{\textit{EFFLUENT DILUTION (BOD)}}$$

On the quantity aspect,[11] one can verify that water users pay for both withdrawal and consumption. The underlying reasoning is that paying for both activities does not constitute double-counting, since withdrawal and consumption rights impose different impacts on other users. Withdrawal rights reduce the availability of consumptive water rights on upstream users. Downstream users are not affected, as water is returned to the water body after use. On the other hand, consumptive rights restrict consumptive and dilution rights both upstream and downstream, once water is permanently unavailable for other uses. Since the water charge formula restricts the withdrawal price multiplier K_0 to be less than one, an indirect relation is established between the importance of consumption and withdrawal. It is assumed that the consumptive use has a bigger impact than withdrawal, since consumption makes water permanently unavailable for other uses. Nonetheless, given that the decision about the value of K_1 is taken by CEIVAP, the 'weight' given to the impacts of withdrawal versus those of consumption is a negotiated decision.

Concerning the water quality aspect, the term $(1-K_2K_3)$ corresponds to a reduction factor applied to the pollution component of the water charge paid by the user. Water users treating larger proportions of their wastewater (implying in a higher K_2) and using more efficient pollution abatement techniques (higher K_3) are given higher reductions. This reduction is intended to be a prime to users that have already been investing in water pollution abatement, and at the same time to act as an incentive mechanism to those who have not engaged in such investments yet. It should be noted that the 'discharge remaining' factor $(1–K_2K_3)$ does not distinguish users by the pollution-intensity of their processes. In this sense, as Formiga-Johnson and Scatasta (2002) observe: 'the dilution factor should be seen as a reward to those who invested in BOD emission reduction, rather than as a way to reflect users' impact on water quality'. Another limitation of the formula is that the water charge only considers BOD, ignoring other pollutants that play an important role in the river due to limitations in the monitoring capacity in this initial phase.

According to the water charge methodology, CEIVAP is in charge of defining the values for the public unit price *PUP* and the withdrawal unit price multiplier K_0. All other coefficients – K_1, K_2 and K_3 – required for the computation of the water charge formula are given by technically defined relations and user-reported information. We now describe the determinants of the water charge criteria.

Public Unit Price (PUP) and Withdrawal User Price Multiplier (K_0)
The determination of the parameters *PUP* and K_0 has taken into account essentially the revenue-generation aspect.

No attempts were made to link the unit price with the economic value of the water. The definition of the value for *PUP* was driven mainly by the need to generate revenue above US$5.45 million, so that the Paraíba do Sul river basin could be eligible to funding by the federal River Basins Cleanup Program (Programa de Despoluição das Bacias Hidrográficas). This governmental programme started in 2001 and aims to support the construction and operation of wastewater treatment plants chosen by River Basin Committees.

In order to assess the revenue-generation potential of the water charges, the Hydrology and Environmental Studies Laboratory of the Graduate School of Engineering, Federal University of Rio de Janeiro (Laboratório de Hidrologia e Estudos Ambientais, COPPE/UFRJ) carried out simulations varying the value of *PUP* from US$7.78/$10^3$m^3 to US$19.46/10^3m^3. In the simulations, only the large users were charged, 'large users' being defined as municipal water services in cities with more than 10000 inhabitants, and the 40 most important water

*Table 5.4   Simulation of the annual revenue-generation results of
              water charges applied to larger users in federal waters in
              the Paraíba do Sul river basin (in US$ millions/year)*

User	PUP = 7.78 US$/$10^3$m^3	PUP = 11.67 US$/$10^3$m^3	PUP = 15.56 US$/$10^3$m^3	PUP = 19.46 US$/$10^3$m^3
Water supply and sanitation services	3.47	5.20	6.94	8.67
Industry	2.12	3.18	4.24	5.30
Total	5.59	8.39	18.11	13.98

Notes: Assumptions: K_0 = 0.5; K_1 = 0.2; K_2 (industrial users) = 1.0; K_2 (domestic)
           including planned investments; K_3 = 0.9

Source: Paraíba do Sul Water Resources Plan (2002).

polluters located in the states of São Paulo, Rio de Janeiro and Minas
Gerais. The results are shown in Table 5.4.

Based on these simulations, CEIVAP approved in March 2001 the
value of *PUP* = R$7.78 /$10^3$m^3 and K_0 = 0.5 for large users. However,
as said before, the universe of payers was considerably enlarged during
the negotiation period of March–December 2001. New simulations
were done in order to assess the revenue-generation potential of water
charges in this enlarged universe of payers, using the values of *PUP* =
R$7.78/$10^3$m^3 and considering the values of 0.4 and 0.5 for K_0. In the
simulations were included the firms responsible for 95 per cent of BOD
emissions and/or with more than 50 employees located in the basin.
Table 5.5 shows the results.

Comparing the results in Tables 5.4 and 5.5, one can observe that
enlarging the universe of payers has little impact in terms of revenue
generation. With *PUP* = US$7.78 /$10^3$m^3 and K_0 = 0.5, the revenue from
charging only large users is US$5.59 million, while the revenue raised
from the enlarged universe is US$5.67 million. So the extension of the
charge to all permit-holders (except non-significant users) was a
decision based more on political and pedagogical concerns, reinforcing
the participatory nature of the new management instrument, than on
financial ones. In December 2001 CEIVAP finally approved the values
PUP = US$7.78/$10^3$m^3 and K_0 = 0.4 to all permit-holders.

Water charge reduction mechanisms   After the approval of the initial

Table 5.5   Simulation of the annual revenue-generation results of
water charges applied to the enlarged universe of payers
in federal waters in the Paraíba do Sul river basin (in US$
millions/year)

User	$K_0 = 0.4$	$K_0 = 0.5$
Water supply and sanitation services	3.39	3.58
Industry	1.82	2.09
Total	5.21	5.6

Notes: Assumptions: PUP = US$7.78/$10^3$m³; $K_0 = 0.5$; $K_1 = 0.2$; K_2 (industrial users) = 1.0; K_2 (domestic) including planned investments; $K_3 = 0.9$.

Source: Paraíba do Sul Water Resources Plan (2002).

water charge proposal in March 2001, industrial users engaged in persistent pressure to add a reduction mechanism into the total water charge value. Talks converged to a global reduction mechanism for both industrial and domestic users. In December 2001 CEIVAP approved the following rules for the reduction factor on the total water charge value:

1. 18 per cent for users paying in the first month of the water charge implementation
2. the reduction factor is decreased by 0.5 per cent for each subsequent month[12]
3. the reduction factor attributed to the users remains valid during the three years of the transitory phase

The main advantage of this reduction mechanism is to offer incentives to users to join the water charge system promptly. Note that negotiation on reductions was based on a political bargain to assure user's commitment rather than on the opportunity cost of revenue's losses.

b)   Agriculture and cattle raising
Excluding the inter-basin transfers, agriculture and cattle raising activities are the main water users in terms of withdrawal and consumption. The Paraíba do Sul river basin includes a total irrigated surface of 123734 ha (2.8 per cent of the total river basin surface), corresponding to a water flow of 49.73 m³/s for withdrawal and 30.28

m³/s for consumption. The main crops are rice (in the São Paulo state region) and sugar cane (in the estuary). The use of inadequate techniques by the sector, such as slash-and-burn agriculture, has contributed to the environmental degradation of the river basin. Cattle breeding is the land use activity which occupies the largest surface in the river basin, with pasture covering 67.4 per cent of the total surface.

Both activities are also major users in terms of effluent dilution, owing to the employment of fertilizers, pesticides and the animal wastes. However, these polluting processes could not be taken into account in designing the water charge, given the lack of data and difficulties in measuring these non-point source pollution processes, preventing any assessment in order to estimate environmental impacts on the river basin.

Agriculture   Cost impacts were the main concern in defining water charges for the agricultural sector. With the intention of measuring the economic impacts of replicating CEIVAP's methodology applied to domestic and industrial users on the agricultural sector, some simulations were conducted. To proceed with the analysis, two crops were selected: rice and sugar cane. These crops occupy the largest surface are in the river basin and, in addition to that, they are the least productive in terms of financial return per irrigation water volume used. Thus if sugar cane and rice growers can bear the water charge, so can the other producers. In the simulation exercise it was assumed that BOD discharge to the water bodies was nil, given the lack of data. This means that only water quantity is charged and then the coefficients K_2 and K_3 are set to be equal to 1 in CEIVAP's formula.

Table 5.6 shows the results of applying the same values defined for industrial and domestic users ($K_0 = 0.4$; $PUP = 7.78/10^3\text{m}^3$) to both crops. As can be verified, the impact in terms of cost increase is substantial. Since profit margins for both crops are low, rice and sugar cane growers would not be able to bear the water charge burden. Given these results, to facilitate the implementation of the water charge in the agricultural sector, a discount factor of 95 per cent was proposed, relative to the *PUP* defined for industrial and domestic users (PUP_{agric} = US$0.39/10^3\text{m}^3$), while keeping the same value for the withdrawal multiplier. The simulation results are given in Table 5.7. As it can be seen, with the new criteria, the impact on costs is less than 1 per cent.

Nevertheless the reluctant behaviour of the agriculture sector towards the water charge against the other user's willingness to have charges

Table 5.6   Economic impacts of water charge introduction in the agricultural sector simulation results

Crop	CEIVAP methodology					Water charge	Cost
	K_0	K_1	K_2	K_3	PUP/10^3m^3	(US\$/year/ha)	increase
Rice	0.4	0.40	1	1	7.78	129.10	17.28%
Sugar cane	0.4	0.39	1	1	7.78	101.44	12.59%

Source: Paraíba do Sul Water Resources Plan (2002).

Table 5.7   Economic impacts of water charge introduction in the agricultural sector simulation results (PUP discounted)

Crop	CEIVAP methodology					Water charge	Cost
	K_0	K_1	K_2	K_3	PUP/10^3m^3	(US\$/year/ha)	increase
Rice	0.4	0.40	1	1	0.39	6.46	0.86%
Sugar cane	0.4	0.39	1	1	0.39	5.07	0.60%

Source: Paraíba do Sul Water Resources Plan (2002).

accepted by all users led CEIVAP to apply even higher discount factors. In November 2002, CEIVAP finally approved the parameter values PUP = US\$0.19/10^3m^3 and K_0 = 0.4 to be applied for irrigation water use. Agricultural water pollution will not be charged (in the CEIVAP formula, this corresponds to set $K_2 = K_3 = 1$), again particularly due to monitoring difficulties. In addition to that, it was defined that the total agricultural water charge cannot exceed 0.5 per cent of the farmers' total production costs. The proposal still has to be ratified by CNRH.

By taking into consideration the question of economic impacts, the formula eases acceptability of the water charges by the agricultural sector. On the other hand, this exclusive focus on economic considerations excludes any concern about resource management and land use.

Cattle raising   The following figures were considered in the estimation of the economic impact of the water charge in the cattle raising sector:

1. withdrawal: 100 l/day for each BEDA unit (animal-equivalent composite unit),[13] i.e., 36.5 m³/year/BEDA
2. consumption coefficient: $K_1 = 0.5$
3. BOD discharge: given the importance of BOD discharge in the cattle breeding sector (especially in confined pork farming) it was proposed to include the effluent dilution component in the water charges for cattle breeders

Like the agricultural sector, a discount factor of 95 per cent on the *PUP* applied to industrial and domestic users was recommended, that is, $PUP = US\$0.39/10^3m^3$.

The simulations shown in Table 5.8 assess the impact of water charges in the sector. Simulation 1 does not consider the effluent dilution use, setting BOD discharges equal to zero. Simulation 2 specifically considers the case of confined pork farming, assuming that the BOD is discharged into water bodies without any treatment (in the CEIVAP formula, this corresponds to set $K_2 = K_3 = 0$). In both simulations, one can see that the water charge impact is sufficiently low to be borne by cattle breeders.

The parameter values adopted by CEIVAP in November 2002 to the animal breeding sector were the same applied to the agricultural use, with $PUP = US\$0.19/10^3m^3$ and $K_0 = 0.4$. The pollution discharge component will be charged only to confined pork farming. In this case, pork farmers must report the wastewater treatment process adopted in order to calculate K_2 and K_3. It was also defined that the total water charge cannot exceed 0.5 per cent of the farmers' total production costs.

c)   Hydropower generation plants
Since hydropower generation does not involve consumptive use or pollution emissions, this activity should be charged just for water withdrawals. However, a particular aspect concerning hydropower plants hinders the application of a pricing formula based on a fixed value per cubic meter. Since the revenue from a fixed amount of water is directly linked to the fall height, a quantity-based criterion would impose low charges for plants with high falls, and high charges for plants with low falls.

Therefore in the case of hydropower generation the CEIVAP adopted a revenue-based criterion that relates water charges to the electricity output levels. Since large-scale hydropower plants are already paying water use charges, CEIVAP's proposition consisted

Table 5.8   Simulation of the impact of water charges in the cattle breeding sector

	CEIVAP formula				Water charge
	K_0	K_1	K_2	K_3	PUP/10^3m^3	(US$/BEDA/year)
Simulation 1	0.4	0.5	1	1	0.39	0.01
Simulation 2	0.4	0.5	0	0	0.39	0.02

Source: Paraíba do Sul Water Resources Plan (2002).

in applying the same methodology to small power plants.

As mentioned before, from the end of the 1980s large-scale hydro-power generation plants were already paying financial compensations for environmental impacts corresponding to 6 per cent of their hydroelectricity gross revenue sales.[14] This percentage was lately revised by Law no. 9984 of July 2000, which created ANA. This law used the term 'water charge' to characterize the financial compensation paid by hydropower plants and increased the payment from 6 to 6.75 per cent, out of which 0.75 per cent is diverted directly to ANA. This fraction is explicitly interpreted in the law as a water use charge. Only small hydropower generation plants are exempted, since the compensation payments are only applied to hydropower plants with a capacity larger than 30 MW.

CEIVAP's proposition consisted in extending the water charge to the small hydropower plants exempted in this federal law, that is, plants producing quantities below 30 MW. In November 2002, CEIVAP approved the water charge replicating the criteria for large-scale plants: water use charge corresponding to 0.75 per cent of the value of electricity produced by small hydropower plants. Total monthly water charge is calculated by the following formula:

$$C = E \times RT \times P$$
   where
C = monthly water charge
E = energy produced (in MWh)
RT = reference energy price (R$/MWh), defined by the National Power Agency (ANEEL)
P = percentage of the value of the electricity produced (defined by CEIVAP to be 0.75 per cent)

Plants producing less than 1 MWh are considered as insignificant users, so they are still exempted of the water charge.

It should be remarked that such criteria do not properly deal with an important role played by hydroelectric plants' reservoirs. By accumulating a water volume in the reservoirs during the rainy season, characterizing an unpaid 'temporary consumption', hydroelectric plants increase water availability during the dry season to downstream users. This positive externality, generated by the river flow regularization role played by hydroelectric plants' reservoirs, should receive a financial compensation in terms of water charge deductions. In the current debate on water charge methodology such a deduction mechanism has not yet been accepted.

d)   Inter-basin transfers

Inter-basin transfer water charges have been the subject of a particular intense debate. In the state of Rio de Janeiro, the main electricity system, Sistema LIGHT, withdraws 160 m³/s from the Paraíba do Sul river, which corresponds to two-thirds of the river flow, for hydropower generation at Lajes Hydroelectric Complex. However, the water is not returned to the original river basin. After being used for hydropower generation, water is transferred to the Guandu river basin, a basin under state domain, that will end up as the main water supply source for the metropolitan area of Rio de Janeiro.

This significant transfer has raised two questions concerning the design of the water charge. The first one is whether use for hydropower generation by the Sistema LIGHT should be considered as consumptive, given that it returns none of the flow it withdraws to the basin of origin. In the debates at CEIVAP, hydropower users have been defending the view that hydropower generation is by nature a non-consumptive activity, and the financial compensation of 0.75 per cent of the value of the electricity produced, already in place, embodies all the environmental impacts produced by hydropower generation activities. On the other hand municipalities, industry and civil organization representatives have expressed the view that, since the water withdrawn is not returned to the Paraíba do Sul river, the consumptive use should not be overshadowed by the one related to hydropower generation.

The second question is whether the burden of charges for the inter-basin consumptive use would have to fall solely onto the hydropower user. The LIGHT Company, owner of the Sistema, maintains that it is not the Sistema that consumes the transferred water, but rather the users in the Guandu river. In fact, LIGHT is responsible for the inter-basin transfer, but users in the Guandu river are beneficiaries of the resulting increase in water availability and, moreover, they are the final consumptive users. These users, which do not incur any costs concerning the inter-basin transfers, include several industries and in particular CEDAE, that is the water utility serving the metropolitan area of Rio de Janeiro. Two approaches about how to treat the consumptive component of the transfers are currently under discussion:

1. The first approach holds that LIGHT Company, being the permit holder for hydropower generation at the origin of the inter-basin transfer, should be viewed as the individual user and be solely responsible for the transfer, being charged for water consumption. This argument faces political obstacles to implementation, since it would imply much higher costs for a sector that is traditionally considered as non-consumptive. Alternative propositions to share the burden of charges, while keeping LIGHT as solely responsible, face constitutional constraints. For example the suggestion that LIGHT should be assigned a consumptive right for the full transfer and then be able to sell its return flows to downstream users is blocked by the Brazilian constitutional disposition which states that water is a publicly owned good.
2. The second approach is based on the argument that the inter-basin transfer should be treated in an integrated way, where the totality of beneficiaries should be interpreted as directly or indirectly responsible for the transfer. This approach assumes the integration between the Paraíba do Sul and Guandu river basins, and it seems to be the preferred one among users and stakeholders involved in the discussions. On the other hand, since this approach involves coordination between federal and state water management institutions, it may take longer to be implemented.

Given the complex arrangements to implement the inter-transfer water charge, CEIVAP decided to extend the negotiation period and inter-basin transfer charges were only implemented in 2004.

e) Other users
Fish farming The small issuance of water permits for fish farming in the Paraíba do Sul river basin indicates that the activity does not have a

significant presence in the region. To estimate the impact of water charges on the sector, simulations were conducted using trout production, the main fish farming activity in the river basin. Assuming an average withdrawal flow of 100 l/s, water consumption and BOD discharge equal to zero,[15] Simulation 1 first applied the criteria designed for industrial and domestic users to the sector. As can be seen in Table 5.9, the impact represents 46.7 per cent of total annual costs. Adopting the same criteria applied to the agricultural sector, Simulation 2 uses a discount factor to the *PUP* in order to imply an economic impact representing less than 1 per cent of cost increase. The resulting *PUP* value is US$0.16/10³m³. In November 2002, CEIVAP approved the criteria suggested by Simulation 2.

Table 5.9   *Simulation of the impact of water charges in the fish breeding sector*

	CEIVAP methodology					Water charge (US$/year)	Cost increase
	K_0	K_1	K_2	K_3	$PUP/10^3\text{m}^3$		
Simulation 1	0.4	0	1	1	7.78	9816.65	46.70%
Simulation 2	0.4	0	1	1	0.16	196.33	0.96%

Source: Paraíba do Sul Water Resources Plan (2002).

Mining Activities   CEIVAP's water charge has limited application to the mining sector since mining activities use mostly groundwater sources that are under the state domain, thus not included in the CEIVAP water charge. The proposed water charge suggests mining activities should be charged according to industrial user rates, that is, for withdrawal, consumption and effluent dilution uses. It was also suggested that the same criteria concerning industrial users should be applied to the sector: *PUP* = US$7.78/10³m³ and $K_0 = 0.4$. However, negotiations at CEIVAP have not reached an agreement about the methodology and the decision about mining water charges was postponed. Like the inter-basin transfers case, CEIVAP have established that mining activities charges must be implemented during 2005.

5.3.3.3   Institutional barriers[16]
As we have previously mentioned, the coexistence of federal and state

waters in the Paraíba do Sul river basin requires a big effort in terms of institutional coordination and pricing criteria harmonization if one intends to design and implement water charges to the whole river basin in an integrated way. The basic question is how to establish common criteria for water charges while respecting the principles of decentralization and stakeholder participation.

The necessary dialogue across institutional frameworks is complicated by their different degree of implementation. While São Paulo and the federal government have considerably advanced in the implementation of the management system and in the discussion about water charge mechanisms, Rio de Janeiro and Minas Gerais are lagging behind. These differences in the actual implementation of the management system prevent any arrangement for the immediate application of water charges for the whole basin. Moreover, without further developments in the implementation of water laws in the states, it is not possible to take into account groundwater management, since groundwater is under state jurisdiction.

The creation of Sub-basin Committees along the Paraeíba do Sul river basin poses the additional problem of coordinating their actions in order not to lose the focus on the global management objectives for the whole basin. This question is particular important in the discussion of the inter-basin water transfer from the Paraíba do Sul river, whose solution depends on coordination with the Guandu river basin committee.

Broadly speaking, the federal and state systems agree about the participatory nature of the river basin committee and its role. However, some differences can be found in the composition of River Basin Committees. In the state of São Paulo, for example, the role of government representatives is much greater than elsewhere, shifting the balance of power away from users and civil society.

Similarly, the four systems define the role of river basin agencies as providing administrative and technical support to committees, with functions that include the collection of water charges and control over their use based on water resource plans. However, there are relevant differences regarding the nature and functioning of water charges. The main question is whether revenue raised by charges should be used within the basin of origin or some degree of 'fiscal solidarity' should be pursued. São Paulo explicitly foresees the possibility of revenue transfers to other basins. This issue has been intensely disputed among water users, state and local interest representatives in the River Basin and Sub-basin Committees.

5.3.3.4   Revenue generation and allocation

The expected revenue to be raised with the introduction of water charges concerning the industrial and domestic users of federal waters is of the order of US$5.21 million/year. This revenue, associated with funding provided by the federal, municipal and state budgets, is supposed to finance the planned investments for the Paraíba do Sul river basin.

The financing decisions were based on the Investment Programs for the states of Rio de Janeiro, São Paulo and Minas Gerais, which identified the main structural and non-structural water-management related measures to be taken in the Paraíba do Sul river basin. Given the complexity of the programmes and the high cost associated with the measures, CEIVAP decided to establish qualitative criteria for the ordering of the measures to be implemented. These criteria were established according to the priorities defined for the Paraíba do Sul river basin.

First of all, CEIVAP considered that water supply is not a critical problem in the river basin, given that almost 100 per cent of the population is connected to municipal water supply networks. The problem of water availability also is not considered critical in the river basin. The most urgent priority for the river basin is water quality control. The following objectives were considered priorities when choosing the funding allocation:

- Objective 1: Implementation of the water resource management system. The implementation of the water management system is a previous condition for the achievement of the other prior objectives. Table 5.10 specifies the main measures and their associated costs for the structuring and operational activities of the water management system. As can be seen, the importance of monitoring is indicated for the magnitude of budget allocations related to this activity.
- Objective 2: Water quality recovery. Water quality is identified as the main problem of the basin. Therefore, pollution control and water quality recovery investments are considered the priority needs for water resources management in the Paraíba do Sul river basin. Domestic and industrial effluents – particularly domestic ones that do not receive any treatment – are the main causes of this poor water quality. The emphasis in pollution control measures are reflected in the total investment figures in Table 5.11 that allocate almost all the budget to financing the construction of wastewater treatment plants and other sanitation works in urban areas.

Table 5.10   Cost estimation of water management and planning measures (in 1000 US$)

Measure	Cost
Implementation of the Paraíba do Sul Water Agency	1970
Implementation of water management instruments and tools	1480
Water information network	220
Technical training programmes	340
Social communication and community mobilization programme	590
Environmental education programme	690
Implementation of automatic monitoring stations	3250
Cartography works	3250
Evaluation of economic and public health benefits	790
Guandu River Basin Water Resources Plan	520
Master Plan on flooding control in the Paraíba do Sul River Basin	690
Total	13790

Source: Paraíba do Sul Water Resources Plan (2002).

Table 5.11   Cost estimation of structural interventions (US$1000)

Measure	Cost
Water sanitation works	28200
Erosion control	1880
Total	30080

Source: Paraíba do Sul Water Resources Plan (2002).

- Objective 3: Erosion control. This objective was set to allocate resources to measures dealing with critical environmental problems of the river basin. Table 5.11 shows that a low percentage of resources is assigned to these interventions, whose measures consist of localized control of erosive processes.

Public perception
Generally speaking, the choice for a simple methodology in defining

the water charge and the participatory nature of the decision-making process has facilitated the acceptability of the water charges.

During the negotiations about the methodology to be applied to the Paraíba do Sul river basin, the industrial and hydroelectric sectors showed strong organizational abilities in order to represent their interests. Both sectors, worried about the economic impact of the charges on their production costs, have participated intensively in the discussions at the River Basin Committee. The adoption of the reduction mechanism for industrial and water municipal services can be viewed as the successful result of the industrial sector's pressure. The electricity sector succeeded in keeping the same percentage stipulated by the federal law for water use charge, 0.75 per cent of the value of electricity produced by hydropower plants, despite of the several proposals to increase this percentage in the Paraíba do Sul river basin. In addition to that, the sector plays an influential role in the development of the negotiations about inter-basin transfers.

Municipal water services, another sector with high organizational capacity, have participated less intensively in the negotiations. Although they have shown concerns about the impact of the charge on consumers' water bills, they have reacted favourably to the introduction of water charges at the proposed levels. Estimates of the impact of water charges on households' water bills have ranged from 1 per cent to 3 per cent, which does not represent a strong impact from the point of view of service providers.

Even with irrigation water representing the largest water use in terms of water withdrawal and consumption (after inter-basin transfers), the sector was not able to organize itself and played a minor role during negotiations. This can be explained by the fact that the sector is atomized and represents a small share of the economy of the basin. On the other hand, the agricultural sector was the most reluctant to accept the water charge mechanism. Since policy-makers were most interested in having all types of water users being charged (and willing to pay), the sector was benefited with very low water charges, despite its weak organizational capacity. The same lack of organizational ability could be observed in the other users (mining, fish farming).

Community and other stakeholders' interests were mainly represented through the activity of environmental NGOs. However, interventions by NGOs are generally localized and isolated, and the lack of financial and human resources has prevented them from playing a systematic role in the discussions.

5.4 FINAL COMMENTS

The use of water charges in Brazil has started very recently and it is still in an experimental phase. The country has, in fact, adopted a very cautious approach and water charges have been introduced in a broad and modern context of integrated water management. A country with the size and hydraulic complexity of Brazil cannot manage water resources through a very centralized and sectoral approach. The first decentralized measure, in place for a long time, differentiates the domain of federal, interstate and intrastate waters.[17] The decentralization process advanced in the 1997 new water law with the adoption of river basin management through River Basin Committees. Institutional arrangements were also tailored with the creation of a Water Resource National Council (CNRH) and an executive federal water agency (ANA) both directly linked to the Ministry of the Environment.

Charges were conceived as instruments to be used along with others, such as permit management plans, to achieve the goals of a water national policy. This policy was intensively debated for two decades and finally approved in the National Congress and State Assemblies.

Nevertheless the aim of applying water charges within the Brazilian legislation seems to be closer to revenue-generation than to efficiency attainment. In addition, critical issues regarding institutional arrangements, universality of uses being charged, autonomy of Basin Committees and charge revenue allocation criteria still dominate the debate and are postponing the rapid implementation of water charges.

Once more Brazil has chosen the experimental path and an important application on water charges is taking place in one of the most important federal river basins, the Paraíba do Sul river that crosses the states of São Paulo, Minas Gerais and Rio de Janeiro. This case constitutes an opportunity to deal with all the issues mentioned above, since it includes conflict with hydroletric generation, inter-basin transfer and the need to accommodate state and federal systems. So far the River Basin Committee of Paraíba do Sul (CEIVAP) is managing to reach consensus within a very participatory process. On the other hand, this consensus seems to have been possible only because of the adoption of a clear revenue-raising approach with a very low budget target and simple rules, in both conceptual and operational terms. This simplicity implies, however, an inability to capture environmental and economic aspects that would make charges more efficient and equitable. The aim is to get users familiar with the water charge system and to learn more about the way their behaviour can be modified. The option for simplicity also makes possible the implementation of the water charge in a short-term horizon, since sophisticated methodologies would require data about water quantity and quality aspects which are not currently available. However, it is too early to evaluate how successful the system will be in gradually introducing efficiency, equity and ecological considerations, and how other applications will benefit from this experience. Whatever the results, by following this piecemeal approach Brazil, as contrasted with other experiences in the region,[18] has been able to avoid the failure of implementing water charges within a weak regulatory framework and institutional capacity.

NOTES

1. This chapter was part of a series of papers commissioned by the Inter-American Development Bank for the Environmental Policy Dialogue and the opinions expressed in this chapter are solely those of the author and do not necessarily reflect the position of the IADB.
2. The authors are very grateful to Francisco Viana and Jair Sarmento from ANA and the research team of the Hydrology and Environmental Studies Laboratory of the Graduate School of Engineering, Federal University of Rio de Janeiro.
3. Before that, the Ministry of Internal Affairs, under which the federal environmental agency SEMA was subordinated, published a resolution (Portaria no. 13) on 15 January 1976 establishing the classification of water in Brazil.
4. Art. 21, inciso XIX, Brazilian Federal Constitution of 1988.
5. Studies on charge proposals are also found for Bahia and Rio do Grande do Sul states, although they are not conveyed through political and institutional channels yet. See detailed description of them in Seroa da Motta (1998) and Asad et al. (1999), including the case of Ceará.
6. This proposal was prepared by a consultancy of the CNEC/FIPE Consortium.
7. Some ideas expressed in this subsection are based on the work of Formiga-Johnson and Scatasta (2002) and on the Paraíba do Sul Water Resources Plan, the latter undertaken by Fundação COPPETEC (2002).
8. Also in implementation are the federal CBHs of Muriaé e Pomba River Basin, São Francisco River Basin, Piracicaba, Capivari e Jundiaí River Basin, Doce River Basin and Paranaíba River Basin.
9. Water volumes are characterized as 'non-significant' for permit-issuance purposes when their withdrawal does not cause a measurable modification of the water resources in terms of quantity and quality.
10. The new management system requires that allocation of water permits be based on priorities defined by water resources plans. Nevertheless, the drive to implement the water charges in 2002 led CEIVAP to adopt a declaratory issuance procedure: users will be given declaratory permits based on their declared historical use.
11. This paragraph is based on Formiga-Johnson and Scatasta (2002).
12. Meaning that if the user delayed payment for a year, the reduction factor would be $18 - 12 \times (0.5) = 12$ per cent.

13. BEDA: animal-equivalent composite unit representing cow, pork and other animal breeding.
14. See page 9.
15. The reason for assuming BOD equal to zero is the difficulty in measuring these quantities, due to the non-source pollution nature of these emissions.
16. This section is based in Formiga-Johnson and Scatasta (2001).
17. Of course, as Brazil is a federation, there is a natural tendency for decentralization.
18. As previously noted for the Colombian case in chapter 2 and described for the Mexican case in chapter 4. In regard to general use of economic instruments and institutional constraints, see Seroa da Motta et al. (1999).

REFERENCES

Asad, M., L.D. Simpson, L.G.T Azevedo, K.E. Kemper, R. Seroa da Motta and M. Bryant (1999), *Brazil: Management of Water Resources – Bulk Water Pricing in Brazil*, World Bank Technical Paper 432, Washington, DC: The World Bank.

Formiga-Johnson, R. and M. Scatasta (2002), 'One Brazil? The impact of regional differences on Brazil's new water management system: an analysis of its implementation in the Paraíba do Sul and Curu river basins', Mimeo.

Fundação COPPETEC (2002), *Plano de Recursos Hídricos para a Fase Inicial da Cobrança na Bacia do Paraíba do Sul*, Rio de Janeiro.

IBGE (2002), *Pesquisa Nacional de Saneamento Básico 2000*, Rio de Janeiro.

PQA (Projeto Qualidade das Águas e Contrôle da Poluição Hídrica) (1999), *Controle da Poluição Hídrica Industrial na Bacia do Rio Paraíba do Sul*, PS-RE-30-RO, PQA/RJ.

Secretaria de Recursos Hídricos (2002), *Evaluation of Brazil's Water*, Brasília.

Seroa da Motta, R. (1998), 'Utilização de critérios econômicos para a valorização da água no Brasil', *Texto para Discussão 556*, Rio de Janeiro: DIPES/IPEA.

6. Conclusions[1]

Ronaldo Seroa da Motta

6.1 MAIN FINDINGS

Next we will discuss the most important issues that carry similarities among the three analysed cases, and the differences in each case, to offer useful insights for improving water management. This analysis is undertaken by phases in which issues related to policy orientation, instrument design and implementation are treated separately.

6.1.1 Policy Phase

Water charges have been introduced within a policy framework
The introduction of water charges (WCs) has occurred within a new policy context. Therefore charges have been considered as instruments to achieve policy goals rather than goals themselves. All three analysed countries were dealing with water policy back in the early nineteenth century. However, increasing water scarcity and environmental problems due to rapid industrialization, urbanization and irrigation have forced policy changes in water resource management. In all three cases, water charges are introduced as instruments for this new water policy approach. This new approach, however, was primarily concerned with the need to plan and decentralize water management in order to accommodate multiple conflicting uses and excesses over assimilative and support capacities of the country's water systems.

The reference experience is undoubtedly the French case, where the 1964 Water Act resulted in new legal and institutional frameworks for water management. The apparent success of this experience was fully absorbed in the Mexican and Brazilian cases.

However, the Mexican pattern has been slightly different. Although it is currently closer to the French approach, use charges were in place from the 1980s without proper institutional and policy frameworks. It was only with the creation of the National Water Commission (CNA) in

1989 and later with the 1992 National Water Law that their implementation was enlarged to pollution matters and conceived as a tool for planning and decentralization.

Water charges are introduced in complement to CAC

Despite the fact that the primary goal of WCs has been nominated to assign an economic value for water, in all cases charges were in place to help the enforcement of CAC instruments, such as permits and standards. That is, no CAC instruments were replaced to give room for a pure economic approach, as it would be the case of Pigovian taxes. For example in the Mexico case, emission standards were revised to accommodate compliance.

Moreover, the new water policy frameworks created new CAC instruments such as the River Basin and National Water Management Plans, where WCs would work to achieve the plans' targets. In fact, these plans happen to be the main instruments in this new policy framework since they merge all the others concerning water availability and priority supply, environmental targets, investment plans and distribution of WC revenues. This will be crucial to analyse implementation issues, as will be developed later, since it will shift the role of WCs to revenue-raising aims from their ability to attain environmental goals.

Decentralization is carried out with river basin institutions

Decentralization is planned in two ways: (1) water management goals and targets differentiated by river basins, and (2) conflicts among users dealt with through a participatory process. Institutional bases for that are the River Basin Committees (RBCs) that define management targets to be executed by their Water Agencies (WAs). This is the basis of the French system in which RBCs take managerial decisions on several water measures, particularly on charge levels.

In the case of Mexico this decentralizing process is less accentuated since the federal water agency, the CNA, is in charge of accommodating the basins' demands and needs, and river basin authorities have been relegated in practice to a secondary role. Brazil has gone further in decentralizing and shifting management power to basin authorities. In Brazil the creation of river basin authorities is not compulsory, and WCs' pricing criteria are defined at basin level, with RBCs consequently gaining more autonomy in this matter than in France and, in particular, than in Mexico.

6.1.2 Design Phase

Water charges are designed as a financing mechanism
Following the same approach adopted in the French system, pricing criteria of WCs take into account assimilative and support capacities of river basins. To accommodate economic and social conflicts they also differentiate by users on sectoral and equity grounds. However, all cases confirm the evidence of MBI literature that WCs are financing mechanisms for investment solutions for water management, including pollution control. This revenue-raising feature is very clear in the Brazilian case, where investment plans, as in the French system, are designed in accordance with WC levels to achieve water management targets. In the case of Mexico, these goals are somewhat obscured by an emphasis on using the WC and exemptions to enforce CAC instruments and targets.

Revenue transfer and exemptions play the major instrumental role
Apart from administrative costs, the major share of WC revenues goes to infrastructure investments and direct transfer for users to finance their pollution abatement actions. Such transfers are thought of as the cornerstone for political acceptance and users' commitment to the charge system. Charge exemptions and rebates are also widely used to protect economic activities, or justified on equity grounds. All this has been pointed out in the French case, as revenue transfer has in fact increased over time, and attempts of the federal government to fund revenues in the general budget have failed. In Brazil the first experience in the Paraíba do Sul river basin has set charge levels according to the financing needs required to leverage federal funds for river clean-up programmes. In Mexico recently the CNA has been explicitly committed to use revenue funds for water-related investments. In all cases, agriculture is either exempted or paying very low charges.

6.1.3 Implementation Phase

Unsolved sectoral conflicts reduce the system efficacy: sectoral conflicts are the main barrier for the full charge application. In France, the charge system was gradually implemented by increasing over time the set of pollutants and sectors. It started charging pollutants that are more easily monitored (industrial and residential organic matters and suspended solids, for example) and from sectors with less political resistance and higher ability to pay (industrial and residential users). It must be noted that ability to pay is used here in the sense of water intensity costs in total operational costs, so the agriculture sector in France was subjected only recently to user charges and is still free of pollution charges.

Mexico, in turn, even adopting the same gradual approach, has failed to fully implement its charge system mostly due to political resistance issues that were not solved prior to the implementation phase. CNA was not able to attract enough federal budgetary means to improve its monitoring and enforcement capacities to collect payments from state-owned sanitation companies and also from several industrial sectors that received waiver schemes during recession periods. All this contributes to undermine the already incipient system and reduce revenue allocation to improve institutional capacity. Mexico has been trapped in this vicious circle despite several modifications in water charge regulation. This can be partly explained by the fact that regulation enforcement in developing countries is generally poor whatever the public policy. But it is also plausible to admit that a greater autonomy of river basin authorities could have mitigated the weak monitoring and enforcement capabilities by accommodating conflicts. The recent movement to a more river basin-oriented approach in this country may change this pattern.

In Brazil, the state-owned hydroelectric companies and the agricultural sector also managed to achieve favourable charge levels. Brazil, however, has adopted a more cautious approach recognizing that the country's territorial and hydrological dimensions would not allow for the immediate creation of a complex structure of river basin management, and has accepted the need to implement it gradually. To achieve this, the new water policy shifts to users the initiative to create a RBC and, therefore, the application of WCs. Only when users fulfil some requirements, such as users' representation, permit regularization and a five-year management plan, is the RBC officially recognized and autonomy for charge application given. That will certainly lead to a slow implementation pattern in the beginning, but it is expected that successful experiences will create incentives for the supply of qualified human resources and the transference of institutional capability that will speed up the whole process of mounting RBCs over the country. Nevertheless the lack of a national grid of River Basin Committees, as in France and in Mexico, poses serious problems related to inter-basin externalities when connected basins are not all organized in RBCs, as already presented in the first major experience of the Paraíba do Sul River Basin Committee.

Participatory process may preclude price incentives

The need of a participatory process to accommodate users' conflicts and to increase acceptance cannot be seen as a sufficient condition to make the most of the potential benefits of a water charge system. The French case has shown that agricultural users can use sectoral subsidies to compensate for the increasing burden of water charges, and therefore to reduce their incentives for changes in water use patterns. It is also known that low charge levels can create incentives for the operation of abatement facilities once they are in place, but it does induce abatement investments that are highly dependent on charge transfer. That is, participation may solve revenue-related conflicts but it does not necessarily create incentives for a charge system that will radically change water pattern uses. In the Brazilian experience of the Paraíba do Sul river basin, charge level setting was initially calibrated according to the minimum economic impact level on users' costs, with no attention given to environmental consequences and water use levels.

Environmental and water management frameworks have to work together

Although monitoring of water use is usually under the responsibility of WAs within the water management framework, water pollution control is exercised by environmental regulators. As said before, in France and in Mexico, where WC systems are already in place, efforts have been made to conciliate the water pollution CAC instruments with the WC systems. However, in both cases joint work in terms of monitoring and information sharing needs to be improved. It is also known that the lack of a continuous evaluation process to analyse the effects of the charge system on use levels and on environment quality has delayed improvements to the system and in the allocation of the WC revenues.

6.2 SUMMARY OF THE RECOMMENDATIONS

Based on the analysis presented above, the following recommendations can be summarized:

6.2.1 Pricing Criteria

A policy framework must be in place before the charges are designed and the charges must be in accordance with policy goals. If revenue-raising goals are the only politically viable option, that should be explicitly acknowledged and the reinforcement of CAC instruments has to be planned.

Even with emphasis on revenue generation, environmental consequences of charge application should be explicitly discussed to allow for gradual incorporation of environmental criteria in the charge system. So continuous environmental evaluation of the river basin should be undertaken, incorporating economic models that identify water use changes related to charge impacts.

Criterion for favourable charge levels should be explicit and based on either economic or equity grounds, and include all users in the charge system to strengthen commitment and enforcement.

6.2.2 Institutional Arrangement

Autonomy of river basin authorities must be tailored according to the dimension and complexity of the hydrological system to maximize institutional capacity by facilitating political acceptance, reducing asymmetry of information and administrative costs.

So the water management framework must be integrated with other policy frameworks to increase monitoring and enforcement capacities. This is the case for environmental agencies as well as sectoral agencies in order to accommodate exogenous policy aims. Since this integration requires federal-level negotiations this is a task for a federal water agency and cannot be delegated to river basin authorities.

6.2.3 Evaluation

It is important to introduce cost–benefit analytical tools to select projects to be financed with charge revenues to maximize social value of the investment actions. And public opinion must be motivated to debate water management issues, with data release and technical arguments to consolidate river basin management and the role of water charges.

NOTE

1. This chapter was part of a series of papers commissioned by the Inter-American Development Bank for the Environmental Policy Dialogue and the opinions expressed in this chapter are solely those of the author and do not necessarily reflect the position of the IADB.

Index

Index